Red Army Self-Propelled Guns
of the Second World War

Red Army Self-Propelled Guns of the Second World War

Photographic History of the Red Army's Second World War Self-Propelled Artillery

Alexey Tarasov

Pen & Sword
MILITARY

First published in Great Britain in 2024 by
Pen & Sword Military
An imprint of Pen & Sword Books Limited
Yorkshire – Philadelphia

Copyright © Alexey Tarasov 2024

ISBN 978 1 39907 151 2

The right of Alexey Tarasov to be identified as
Author of this Work has been asserted by him in accordance
with the Copyright, Designs and Patents Act 1988.

A CIP catalogue record for this book is
available from the British Library

All rights reserved. No part of this book may be reproduced or
transmitted in any form or by any means, electronic or mechanical
including photocopying, recording or by any information storage and
retrieval system, without permission from the Publisher in writing.

Typeset by Mac Style
Printed in the UK by CPI Group (UK) Ltd, Croydon, CR0 4YY.

Pen & Sword Books Limited incorporates the imprints of After the Battle,
Atlas, Archaeology, Aviation, Discovery, Family History, Fiction, History,
Maritime, Military, Military Classics, Politics, Select, Transport, True Crime,
Air World, Frontline Publishing, Leo Cooper, Remember When, Seaforth
Publishing, The Praetorian Press, Wharncliffe Local History, Wharncliffe
Transport, Wharncliffe True Crime and White Owl.

For a complete list of Pen & Sword titles please contact

PEN & SWORD BOOKS LIMITED
47 Church Street, Barnsley, South Yorkshire, S70 2AS, England
E-mail: enquiries@pen-and-sword.co.uk
Website: www.pen-and-sword.co.uk
or
PEN AND SWORD BOOKS
1950 Lawrence Road, Havertown, PA 19083, USA
E-mail: uspen-and-sword@casematepublishers.com
Website: www.penandswordbooks.com

Contents

Acknowledgements   viii
Introduction   ix

Chapter 1   Lost Opportunities: Development and Production   1

Chapter 2   Organization and Order of Battle   12

Chapter 3   Training: Sweat Saves Blood   21

Chapter 4   Tactics and Doctrine   32

Chapter 5   In the Flames of Battle   45

Chapter 6   Allied Assistance   59

Conclusion   The Perfect Combination   66

The Photographs   75

Interwar   77
SU-18 self-propelled gun (drawing)   77
SU-2 self-propelled gun   78
SU-1 self-propelled gun   79
AT-1 artillery tank   80
SU-5 "small triplex"   82
SU-14   86
T-27   89
T-27M   90
SU-37 self-propelled anti-tank gun   91
SU-76 (T-27-based)   91
SU-6 self-propelled AA gun   93
SU-12   94
29K truck-borne AA gun system   95

1941–1943	96
ZiS-30	96
ZiS-41	97
ZiS-43	101
Armored ZiS-5	103
Armored GAZ-AA	104
Object 212	104
KV-7	105
SU-71	106
SU-76 (SU-12) M1942	108
SU-76M (SU-15)	109
SU-76BM (NATI-TsAKB)	113
SU-38	114
SU-74B (SU-57B) M1943	116
SU-74D (SU-76D) M1943	117
SU-85A (SU-15A) M1943	118
SU-122	120
SG-122 (A)	130
SU-85	131
SU-76i	147
SU-152 (KV-14)	149
S-51 SPG (Autumn 1943)	153
SU-11 SPAAG	157
SU-31	158
SU-37 / ZSU-37	161
T-70-based 37mm AA gun (ZUT-37 early)	172
T-70-based twin MG mount	173
ZUT-37 SPAAG (late)	174
1944–1945	176
SU-85B	176
GAZ-75	178
SU-57	182
SU-122P	184
KSP-76 (GAZ-68)	188
OSU-76	189
SU-100 (Object 138)	192
SU-101 and SU-102	196
ISU-122	200
ISU-152	207
ISU-130 (Object 250)	211
Kirovets-2 (Object 704)	213

Lend-Lease	214
M10	214
M15	214
M17	215
T48	216
Notes	217

Acknowledgements

This publication would not have been possible without the help and support of some close friends and my family.

I extend my greatest thanks to Cyrill Kopylov and Sergey Makarov for providing materials and valuable insights into my research.

Special thanks to my wife Maria for her continuous support and assistance in this journey.

Introduction

The story of Russian self-propelled artillery begins in the early 20th century. Since then, it has undergone significant development and now holds a crucial place in the structure of the modern Russian Army.

However, the evolution of this arm of service was neither easy nor linear. Along its path, it experienced numerous highs and lows, closely tied to political and social changes in Russia that inevitably influenced the development of its armed forces. After a decline caused by the Russian Revolution and the Civil War, self-propelled artillery was reborn and flourished during the 1930s, aligning with the ideas of the Deep Battle doctrine and the extensive growth and technical rearmament of the Red Army.

These developments were disrupted with the beginning of the Great Purge. During this time, many talented engineers, theoreticians, and military commanders, including the evangelists of the innovative Deep Battle doctrine, were accused of treason and arrested or executed.

The evolution of self-propelled artillery essentially stalled until the beginning of the Great Patriotic War. By 1942, the Soviet leadership recognized the necessity for reintroducing self-propelled artillery into service. This was done in 1943, and since that moment, the new arm of service experienced steady numerical growth and continuous technical improvements. However, other aspects, such as organizational structure or the theoretical understanding of how to employ self-propelled artillery alone or in combined arms warfare, had to be rebuilt from scratch. This led to frequent changes and adjustments that affected the performance of self-propelled artillery on the battlefield.

The Red Army emerged victorious from World War II, but many imperfect decisions, including the structure of the Tank and Mechanized forces and self-propelled artillery, remained in place until the 1960s.

This book investigates the dynamic evolution of Soviet self-propelled artillery and its significant transformations during the period 1930–45. Due to the enormity of the topic, certain aspects, like Soviet self-propelled rocket artillery, have been deliberately omitted. Other elements, such as the development of Russian self-propelled artillery between 1900 and 1940, are discussed briefly. The resulting research primarily focuses on the development of tube self-propelled artillery in the period between 1941 and 1945.

Perhaps the main distinctive feature of this book is that it does not focus solely on technical aspects, design descriptions, and production. This research goes beyond pure technology, aiming to depict self-propelled artillery as a cohesive system, which includes doctrine, materiel and production, organizational structure, training, and, ultimately, tactics.

Each of these aspects is investigated in a separate chapter, culminating in a section describing combat employment from the introduction of self-propelled artillery in early 1943 to the fierce battles in Prussia and the Battle for Berlin in 1945. This section also includes examples of combat episodes, enabling readers to examine how the entire system worked on the battlefield.

An entire chapter is dedicated to the self-propelled artillery supplied to the Soviet Union through the Lend-Lease agreement. It played a significant role in the Great Patriotic War and had an impact on the post-war development of the Soviet tank forces.

The second part of the book features a vast collection of illustrations covering the story of Soviet self-propelled artillery from the 1930s to 1946. Each photograph is accompanied by an extensive caption, providing insights into each model, its modifications, distinctive features, and ultimate fate. Although some prototypes and "paper projects" were omitted, the author hopes that this approach will make it easier for readers to understand the entire "genealogical tree" of this class of armored fighting vehicles.

Finally, the book includes statistical data, tables, and schemes that provide a broad perspective on the substantial aspects and processes of the entire Soviet military system. It's noteworthy that the research is based on archive documents and war diaries of Soviet tank and self-propelled artillery units, many of which have not been published before.

Chapter 1

Lost Opportunities: Development and Production

The Russian Army has maintained a strong emphasis on the artillery branch since its introduction in the 14th century. Throughout the ages, artillery, recognized as a crucial capability, has received substantial investments and undergone steady technological evolution, enabling it to remain one of the most modern and effective branches of the Russian Army. By the 19th century, the Russian artillery corps was a well-established branch of service, relying on a professional officer corps, which included many talented engineers and inventors. It often served as a catalyst for technological innovations.

At the beginning of the 20th century, the Russian Imperial Army (RIA) embarked on its journey towards modernization and motorization, closely monitoring tactical and technological advancements in other countries. Between 1905 and 1912, a large-scale military reform was initiated to restructure the army and introduce new armaments, methods, and forms of combat. Naturally, the reform also encompassed the artillery. For instance, in 1910, new 122mm and 152mm howitzers, along with the Maxim machine gun and quick-firing guns, were adopted into service. In the same year, the first automotive company (later known as the Training Automotive Company) was formed.

Unsurprisingly, the first projects of self-propelled guns were introduced and published during this period. One of the pioneers in this new field was artillery officer Vasily Tarnovsky, who proposed and published a project for a 76.2mm anti-aircraft gun on an automotive chassis in 1912.[1]

The emergence of self-propelled artillery was inevitable, and indeed, the first self-propelled guns and artillery units were deployed to the battlefield in the early months of the Great War. Remarkably, self-propelled guns, in the modern understanding of this class of combat vehicles, were adopted by the Russian Army before tanks.

The construction of the first Russian self-propelled gun began in August 1914 at Izhora Plants. The works were authorized by the Minister of War of the Russian Empire, Sukhomlinov, and were part of a larger undertaking to establish the "armored machine gun car battery"—essentially, the first armored car unit in the Russian Imperial Army.

The design of the combat vehicle was offered by Colonel Alexander Dobrzhansky, a distinguished engineer and pioneer in armored cars. It was built on a Mannesmann-Mulag 5-ton truck chassis, armed with a 47mm Hotchkiss naval gun and two machine guns. This combat vehicle, along with two other truck-based armored cars armed with Maxim-Nordenfelt 37mm guns, was included in the ranks of the 1st Automobile Machine Gun Company, which was dispatched to the front on October 19, 1914.

Shortly after, another self-propelled artillery unit followed. In late March of 1915, the four-gun battery, known as the "separate Automotive Battery for Shooting at the Air Fleet" and led by Vasily Tarnovsky was sent to the theater of operations.[2] This unit, also known as Tarnovsky's Battery, was the first Russian unit dedicated to air defense and armed with self-propelled anti-aircraft guns (SPAAG).

The first combat deployment of self-propelled guns from the 1st Automobile Machine Gun Company occurred on November 9, 1914, where the gun armored cars proved effective in the close infantry support role. From this moment onward, the armored branch of the Russian Imperial Army experienced steady growth with new models developed and introduced into service. Arguably, the most successful among them was the Garford-Putilov heavy armored car armed with a 76.2mm gun, which was often employed as an assault gun.

One of the distinguishing features of the emerging armored branch was emphasis on wheeled armored vehicles. Generally, this focus was predetermined by two factors. Firstly, the geography of the Eastern Front, with its vast territory and more maneuverable character of warfare in comparison with the Western Front. Secondly, the availability and technical characteristics of the base chassis produced or assembled from imported components in Russia. For example, Tarnovsky's SPAAGs were constructed on the basis of 5-ton trucks built at the Russo-Balt plant in Riga.

In parallel, theoretical and experimental works on self-propelled artillery were underway. These efforts led to the emergence of interesting concepts like Filatov's tricycle armored car. Armed with machine guns or 76.2mm M1910 guns, these vehicles were intended for maneuver groups that could rapidly deploy on a likely axis of the enemy's advance or reinforce defensive lines. Essentially, these vehicles were light self-propelled guns, and their combat use resembled the recommendations for tactical employment of self-propelled artillery in defense that emerged during the Great Patriotic War.

Further experiments envisioned, among other projects, the development of a 107mm self-propelled gun on an all-wheel-drive chassis. However, the Russian Revolution, followed by political turmoil and the beginning of the Civil War, postponed these plans.

Nevertheless, the first large-caliber self-propelled guns were constructed during the Civil War. These improvised combat vehicles were built on the basis of armored Bullock-Lombard tractors and armed with 120mm Canet naval guns. At least two of these vehicles served in the 6th Tractor Battalion (division) of the heavy naval artillery of the Caucasian Army.

It is worth noting that the change in the political regime in Russia did not interrupt theoretical works on artillery. Old institutions, such as the Main Artillery Directorate (GAU), continued their work and retained a substantial part of their old staff.

In December 1918, the Artillery Committee of the Main Artillery Directorate established a special commission for special artillery experiments (KOSARTOP). This commission was led by Vasily Trofimov, a Lieutenant General of the RIA, expert on ballistics, and designer of artillery systems.

In 1920, under the leadership of another former Lieutenant General of the RIA and prominent artillery designer, Rostislav Durlyakhov, a "Commission for the Military Industry Program" decided to initiate works aimed at the development of self-propelled guns for direct infantry support.

An important point to consider is that the young Red Army inherited not only institutions and a vast surplus of armaments from the Imperial Army, but also access to knowledge and expertise in many fields crucial for further development. Many officers from the old army who decided to stay in Soviet Russia served as the so-called military specialists or "voenspetsy." In total, over 75,000, or 30% of the old officer corps, were involved in service for the RKKA.[3] These individuals, possessing the necessary knowledge, skills, and energy, contributed to maintaining continuity in the development of many theoretical ideas conceived before and during World War I.

However, the vast majority of experimental projects on self-propelled artillery did not progress beyond the blueprints with only few of them reaching the prototype stage. The main reason for this was the general decline of the Red Army between the end of the Civil War in 1922 and 1926. Until the late 1920s, the Red Army's armored branch primarily comprised of combat vehicles inherited from the RIA or captured from the White Army during the Civil War. Furthermore, the military industry was in a deplorable state compared to the level of 1916–17, impeding the development of technical branches.

The major changes were initiated following Mikhail Tukhachevsky's report, "The Defense of the USSR," during the Executive Meeting of the Council of Labor and Defense in 1926. In this report, he explicitly stated that "neither the Red Army nor the country is ready for war."[4] This event, followed by a period of war scare, prompted forced industrialization and the beginning of a massive army reform.

Importantly, the reform was largely based on the Deep Battle doctrine and focused on the development of technical branches of service. In 1929, for the first time in history, the system of "Tank-Tractor-Automotive-Armored armaments of the Red Army" was outlined. This early system included four types of self-propelled guns dedicated to different combat roles: an anti-tank tankette armed with a 37mm gun; a 76mm fire support self-propelled gun (SPG); and two self-propelled anti-aircraft guns—one armed with a quad 7.62mm machine-gun mount and the other with a twin 37mm anti-aircraft gun mount.[5]

More importantly, by that time self-propelled artillery was recognized as a crucial and integral component of independent mechanized formations of the highest level, such as corps and divisions. This perspective was reflected in doctrinal documents of the highest order. Vladimir Triandafillov wrote in his report to Kliment Voroshilov: "The artillery of mechanized formations should be capable of accompanying them with fire and maneuver. Only under these conditions could it, with its speed, mobility, and readiness to open fire, align with the tactical spirit of mechanized units."

He also believed that, compared to self-propelled artillery, towed artillery would always be inferior in terms of mobility and readiness to open fire in any direction without lengthy preparation. Mobility and cross-country capability were considered key aspects. According to the doctrine, all components of mechanized units should possess uniform technical and tactical characteristics, enabling them to move at a consistent speed and, consequently, act as a single whole.[6]

However, early theoretical considerations possessed some shortcomings. For example, Triandafillov believed that combat actions involving mechanized formations would be characterized by great speed of advance and, accordingly, would rarely encounter resistance from an entrenched enemy. Therefore, he considered the development of indirect fire howitzers unnecessary. Instead, Triandafillov proposed designing a "universal" gun system capable of firing at both flat and high trajectories with reduced charges.[7] It is worth emphasizing that the concept of "universal" dual-purpose self-propelled artillery, capable of providing both direct and indirect fire support, would later become an integral element in the design of combat vehicles such as the SU-152 or SU-122.

The adoption of the T-26 (Vickers 6-ton) tank and the T-27 tankette (licensed Carden Loyd tankette) in 1931 streamlined experimental works and the production of armored combat vehicles, including self-propelled artillery. Another positive contributing factor was the rapid evolution of the Soviet industry, which now enabled the production of complex systems.

The availability of robust, mass-produced combat vehicles enabled the design of self-propelled guns on various base platforms and in different weight categories. As a result, the first half of the 1930s became a period of extensive growth for self-propelled artillery, with essentially all classes and types of these combat vehicles being in development or production. Some of these designs were adopted into service, underwent limited production, and saw action in local conflicts during the late 1930s or in the opening phase of the Great Patriotic War.

Importantly, the evolution of Soviet armed forces was systematic, holistic and proceeded within the framework of a unified doctrine.[8] The doctrinal thought was not inflexible and promptly adapted to changes. For instance, the perspective about the irrelevance of self-propelled howitzer was deemed erroneous and abandoned.

In April 1933, Chief of the Main Artillery Directorate Efimov presented a report on the revision of the system of artillery armaments. This report outlined the system of self-propelled artillery for independent mechanized formations, which included "a) 45mm universal automatic gun; b) 76mm gun; c) 122mm howitzer (on a common self-propelled chassis); d) 152mm mortar; d) 76mm anti-aircraft gun on a self-propelled chassis; e) special power artillery of the High Command Reserve (ARGK) on a self-propelled carriage."[9] The necessary changes were accepted in the same year and included in Voroshilov's report presented in August 1933. The report had two explicitly formulated tasks for the Second Five-Year Plan: firstly, to equip mechanized and tank forces with specialized self-propelled artillery, and secondly, to fulfill the tasks of strengthening air defense (PVO) and anti-tank defense (PTO) for both frontline and rear areas.[10] These plans spurred active developments in the class of indirect fire self-propelled guns, with attempts to design families of SPGs on unified platforms with different armaments. Notable projects in this endeavor included the SU-5 family of vehicles, "The Small Triplex," and SU-7 or "The Big Triplex."

Eventually, the systematic approach led to the establishment of a self-propelled artillery system, comprising four main classes:

- Self-propelled assault guns for direct fire support;
- Self-propelled howitzers or mortars for indirect fire support;

- Self-propelled anti-tank guns or tank destroyers;
- Self-propelled anti-aircraft guns.

These major classes were further subdivided based on weight (light, medium, and heavy) and platform (tracked, wheeled, half-tracked) subclasses.

The steady evolution of Soviet mechanized forces came to a halt in 1936–37 due to the unfolding events of Stalin's Great Purge. On a broader scale, the Great Purge had a negative influence on every aspect of the Soviet military and industry. This impact manifested in disorganized production and a significant setback in the theoretical thinking and general understanding of the role and structure of mechanized forces.[11] As a consequence, numerous self-propelled artillery projects were either cancelled or indefinitely postponed. Additionally, vehicles already accepted into service were withdrawn from serial production.

While some prototypes saw action in the Battles of Khalkhin Gol and during the Winter War, the development of self-propelled artillery was largely halted. However, the experience of the Soviet-Finnish war and the German campaign in Europe raised concerns within the Soviet command and forced revisiting this matter.

In May 1941, Grigory Kulik, Marshal of the Soviet Union and Chief of the Main Artillery Directorate, wrote a memorandum to Stalin and Zhdanov, stating that the armament system of the Red Army lacked any self-propelled artillery. He stressed that "all works in this direction conducted between 1930 to 1935 were discontinued, and until that moment, this class of weapons had received no attention." Furthermore, he referenced the experience of the Soviet-Finnish war and presented arguments on the necessity to include four classes of self-propelled artillery into the armament system.

Unsurprisingly, these were a 76mm assault gun for supporting mechanized and motorized infantry units; a heavy 152mm gun for fighting against fortified positions; 57mm and 85mm anti-tank guns; and a 37mm anti-aircraft gun. In addition, Kulik offered to design APCs for mechanized infantry and armored ammunition transporters. In other words, the Soviet command returned to the ideas previously formulated by the Deep Battle evangelists, Tukhachevsky, Triandafillov, Kalinovsky and others.[12]

While Kulik's memorandum was sound, the beginning of the Great Patriotic War in June 1941 made this matter irrelevant. Following a series of defeats in the first months after the invasion and the rapid advance of the German forces, by October 1941, the Red Army faced shortages in basic equipment and was not in shape for adopting a new class of armament. In addition, Soviet industry struggled after losing industrial areas and resources in the west of the country, facing chaos caused by evacuation.

It is significant to note that there were research and development projects of self-propelled guns in 1940–41. However, these projects were at early stages or purely theoretical by the beginning of the war.

Attempts to build self-propelled artillery on existing platforms were made in 1941 and early 1942, resulting in limited production of makeshift armored tractors or self-propelled guns, such as the ZiS-30. Interestingly, following the German invasion, the Soviet leadership reverted to pre-war ideas that were heavily criticized in the mid-1930s. In July 1941, the

State Defense Committee issued a resolution #219cc, ordering the construction of 2,000 armored tractors armed with a 45mm gun at the Kharkov Tractor Plant (KhTZ) in the Q3 1941.[13] These plans never materialized, as the Red Army suffered a defeat near Kiev. On September 12–13, 1941, the decision was made to evacuate the Kharkov Industrial cluster to the East.

The Soviet leadership revisited the issue of self-propelled artillery in 1942, once the operational and strategic situation had stabilized. On April 15, 1942, the plenary session of the Artillery Committee of the Main Artillery Directorate approved the development of self-propelled guns armed with a 76.2mm ZiS-3 gun or a 122mm M-30 gun for infantry support, a 152mm ML-20 gun for destroying enemy fortifications, and a 37mm anti-aircraft self-propelled gun.[14]

Prototypes of the light SU-12 (SU-76) and medium SU-122 self-propelled guns were developed and tested from May to December 1942. In early December, these vehicles were approved for serial production, and the initial production batch was ready in early January 1943.

The heavy SU-152 self-propelled gun was developed in early 1943. The project was initiated on January 4, 1943, while the prototype was ready in only 20 days, by January 24, 1943. After testing, the vehicle was accepted and put into serial production on February 14, 1943.

To save time and resources, all self-propelled guns were designed around vehicles and systems that were already in serial production. The SU-76 used components and systems from the T-60 and T-70 light tanks; the SU-122, SU-85, and SU-100 were based on the T-34 tank; while heavy SPGs such as the SU-152, ISU-152, and ISU-122 were built on the KV-1S and IS heavy tank chassis.

The rare exception was the SU-76I SPG built on a basis of captured German Panzer III tanks or Sturmgeschütz III SPGs. On January 19, 1943, the tactical-technical requirements for the 76mm SU-76 assault gun were issued. The SU-76I was designed to provide fire support for tank and infantry units and was intended to be capable of engaging enemy machine gun emplacements, tanks, and infantry using both direct and indirect fire.[15] Between March and November 1943, a small batch of 201 vehicles was produced at the Factory #37.

Production centers for self-propelled artillery were distributed accordingly, depending on where the base platforms or main components were manufactured. Light self-propelled guns were manufactured at Factory #38 (Kirov), Factory #40 (Mytishchi), and GAZ in Gorky (Nizhny Novgorod). The production of medium SPGs was concentrated at the UZTM plant in Sverdlovsk (Yekaterinburg). Finally, production lines for heavy self-propelled artillery were established in Chelyabinsk at the Chelyabinsk Kirov Plant (ChKZ), also known as "Tankograd."

Starting from 1943, the production of self-propelled guns experienced stable growth, reaching its peak in 1944 with a total of 12,060 SPGs produced. The total production numbers are presented in the tables below. Note that the numbers vary depending on the source due to different accounting methods. For example, A. Solyankin, M. Pavlov, I.

Pavlov, and I. Zheltov provide total numbers of SPGs produced between July 1941 and June 1, 1945, whereas other authors cover the entire six months of 1945 in their calculations, resulting in slightly higher totals of 22,436 (Melnikov) and 22,411 (A. Ermolov) SPGs produced. Additionally, some authors exclude prototypes or low-rate production batches from the final numbers.[16, 17]

Table 1. Annual production of self-propelled guns categorized by type from July 1, 1941, to June 1, 1945.[18]

Model	Manufacturer	Year					Total
		1941	1942	1943	1944	1945	
ZiS-30	Factory #92	101	0	0	0	0	101
SU-76	Factory #38	0	25	535	0	0	560
SU-76M	Factory #38	0	0	562	1103	0	1665
SU-76M	GAZ	0	0	601	4708	2214	7523
SU-76M	Factory #40	0	0	210	1344	752	2306
SU-76I	Factory #37	0	0	201	0	0	201
SG-122	Factory #592	0	9	12	0	0	21
SU-122	UZTM	0	26	612	0	0	638
SU-85	UZTM	0	0	761	1578	0	2339
SU-100	UZTM	0	0	0	500	1060	1560
SU-85M	UZTM	0	0	0	315	0	315
SU-152	ChKZ	0	0	668	2	0	670
ISU-152	ChKZ	0	0	35	1340	510	1885
ISU-122	ChKZ	0	0	0	945	490	1435
ISU-122S	ChKZ	0	0	0	225	250	475
Total		101	60	4197	12,060	5276	21,694

Table 2. Self-propelled guns production distributed by weight class per year.[19]

	1941	1942	1943	1944	1945	Total
Light	101	34	2121	7155	2966	12,377
Medium	0	26	1373	2393	1060	4852
Heavy	0	0	703	2512	1250	4465
Total	101	60	4197	12,060	5276	21,694
Light	100%	57%	51%	59%	56%	57%
Medium	0%	43%	33%	20%	20%	22%
Heavy	0%	0%	17%	21%	24%	21%

Table 3. Production of self-propelled guns categorized by type on October 1, 1945.[20]

Type	Produced	Produced, %	Production Start	Production End
ISU-152	2275	9.0%	Dec 43	–
ISU-122	2410	9.5%	Apr 44	–
SU-152	670	2.6%	Feb 43	Nov 43
SU-100	2335	9.2%	Sep 44	–
SU-122	640	2.5%	Dec 42	Sep 43
SU-85	2650	10.4%	Aug 43	Nov 44
SG-122	21	0.1%	Oct 42	Jan 43
SU-76	14,152	55.8%	Dec 42	–
SU-76I	200	0.8%	Apr 43	Dec 43
SU-37	12	0.0%	Feb 45	Apr 45
Total	25,365	100%		

The production numbers clearly demonstrate that Soviet industry successfully achieved its primary goal of saturating the Tank and Mechanized Forces with self-propelled guns, providing an adequate quantity for establishing and equipping new units and replacing lost equipment. However, it is significant to highlight that the foundation for the industrial successes of the 1940s was laid in the 1930s when the decision to move towards the mechanization and motorization of the Red Army was made.

Other factors that contributed to sustaining mass production at this scale and pace were the centralized management and planning, coupled with the ability of the Soviet system to accumulate resources. On the other hand, Soviet industry, in many cases, suffered from a lack of competent administrators, resulting in inefficiency and the wastage of resources.

In terms of design, both self-propelled artillery and the tank industry underwent continuous modernization during the war. Designers aimed to enhance the combat capabilities of self-propelled guns, while factories focused on simplifying production processes to increase output. While both of these goals were largely achieved, the most effective modifications and combat vehicles were designed and entered serial production in late 1943 and 1944. Consequently, many of them only reached the battlefield during the closing phase of the war.

There were also negative aspects that affected the Red Army and the Soviet tank industry as interrelated parts of the same military-industrial system. Firstly, the army simultaneously experienced extensive growth of the Tank and Mechanized Forces and suffered heavy losses in materiel. In addition, the nature of warfare on the Eastern Front was not static and underwent significant changes over the course of the Great Patriotic War. These factors led to the necessity of maintaining a high production pace sufficient to continuously replenish losses, form new units, and introduce new combat vehicles.

These requirements were hardly achievable and tightly bound the factories to simpler designs, components, and existing combat vehicles that were serially produced, as there was practically no time for research and development or the adoption of more complex systems. Moreover, the factories (or plants) themselves were not inclined toward changes, as it meant launching new production processes with inevitable errors, delays in the schedule, and ultimately a disruption of the plan.

This, in turn, resulted in persisting technical deficiencies in armored fighting vehicle (AFV) designs. Similar to larger systems and organizational structures, a faulty part could decrease the efficiency of the entire system. This principle applied to the AFV design resulted in inadequate components causing a decrease in their overall combat capabilities.

For instance, a representative example was provided in a report written by Lieutenant Colonel-Engineer Kostsov after the inspection of self-propelled artillery units in the autumn of 1943. The report stated, "The situational awareness of self-propelled guns is inadequate; the vehicles are half-blind. During combat actions, commanders have to open hatches to find enemy targets or place observers on top of the SPGs tasked to spot and acquire targets and adjust fire. This method results in high losses in men and equipment of self-propelled artillery. It is desirable to design self-propelled guns similar to American SPGs."[21] As demonstrated by this example, despite having sufficient armor, adequate mobility, and capable armament, the self-propelled guns had limited combat value due to being "half-blind" on the battlefield.

Such issues, as described above, were common in Soviet SPG designs, and most of them persisted throughout the war. The most acute problems included ergonomics, situational awareness, optics, radio equipment, and components of the running gear, such as gearboxes.

The second aspect was imbalance in the structure of self-propelled artillery and more widely in the system of armaments of the Tank and Mechanized Forces. Typically, the military, based on the spectrum of the most probable threats and forms of armed conflict, develops a doctrine. This doctrine, in turn, shapes the structure of armed forces and the armament system, down to individual components. In the case of Soviet self-propelled artillery, it was reintroduced to the Red Army after the period of organizational and doctrinal disorder of the late 1930s and in the middle of the war. In other words, the armament system of the Tank and Mechanized Forces, including self-propelled artillery, was not formed based on a fully-fledged doctrine. Instead, it had to be continuously adjusted to the available resources and the actual capabilities of the Red Army. An unbalanced system eventually materialized, where one component, the doctrine, underwent constant changes adapting to emerging battlefield conditions, while the other component—equipment—remained virtually static until the end of the war.

This imbalance of the system can be perfectly illustrated by the following example. The SU-122 self-propelled gun was accepted into service in December 1942, but was removed from serial production in August-September 1943. The main reason for ending production of the SU-122 and prioritizing the SU-85 was the introduction and proliferation of new heavily armored German tanks, such as the Panzerkampfwagen VI Tiger and Panzerkampfwagen V Panther.

However, while the SU-122 and SU-85 shared many technical similarities, they were intended for completely different tactical roles. The SU-85 was a tank destroyer, whereas the SU-122, equipped with a 122mm howitzer, was technically and tactically closer to assault guns and indirect fire howitzers intended for direct infantry support.

Moreover, the SU-122 proved extremely useful on the battlefield, as revealed by archive documents. Soviet officers valued SU-122 SPGs and expressed regrets when their production was canceled.[22] In particular, they noted excellent mobility and reliability, characterizing the SU-122 as "the most maneuverable and suitable for actions with mechanized forces." On the contrary, SU-152 heavy assault guns that remained in production longer were described as "overly heavy with low mobility," posing difficulties in the redeployment of tank and mechanized units.[23]

After the withdrawal of the SU-122 from serial production, the Red Army did not receive any replacement self-propelled howitzers, despite the continuously growing need for this class of self-propelled artillery during the war. At the same time, the Soviet industry continued the production of SU-76 SPGs, despite the fact that by late 1944, these guns were considered incapable for many combat missions and limitedly effective. In the end, SU-76 SPGs accounted for over 55% of the self-propelled guns produced by Soviet industry.

It is remarkable that the accuracy of pre-war developments in self-propelled artillery was repeatedly confirmed during and after the war. In 1943, Lieutenant Colonel-Engineer Demyanenko visited the United States to study the self-propelled artillery of the U.S. Army. In his report, he outlined the American self-propelled artillery system, essentially validating the systems developed in the USSR from 1929 to 1940. He identified the same basic classes:

- Self-propelled anti-tank guns;
- Self-propelled anti-aircraft guns;
- Self-propelled for infantry fire support, guns and howitzers;
- Tank destroyers;
- Tanks-destroyers.[24, 25]

The major differences in the American system included a wider use of wheeled and half-tracked platforms and a much stronger emphasis on self-propelled howitzers.

After the war, in July 1945, Major General of the Engineering and Tank Service Alymov wrote a report on the prospects for the post-war development of self-propelled artillery in the Red Army. Essentially, he reiterated all the systems outlined between 1929 and 1945 with some modifications. He proposed a system with three main classes of self-propelled artillery:

- Assault self-propelled artillery;
- Self-propelled anti-aircraft guns;
- Howitzer self-propelled artillery.

The fourth class, anti-tank self-propelled artillery, was combined with assault guns into a single class. This innovation was likely based on wartime experience, where self-propelled artillery was employed in both roles. Another significant change was a strong emphasis on self-propelled artillery equipped with howitzers.[26]

Table 4. The system of self-propelled artillery proposed by Major General of the Engineering and Tank Service Alymov, July 25, 1945.[27]

Model	Class	Role
SU-100	Assault	Anti-tank, infantry and cavalry support
SU-122	Howitzer	Counterbattery fire, fighting against fortifications, fire support
SU-122BM	Assault	Fighting against fortifications with direct fire
SU-152	Howitzer	Counterbattery fire, fighting against fortifications, fire support
SU-152BM	Assault	Anti-tank, fighting against fortifications
SU-203	Howitzer	Counterbattery fire, fighting against fortifications, fire support
N/A	SPAAG	Air defense
N/A	SPAAG	Air defence

Essentially, in 1945, Alymov was on the brink of inventing the system that would later emerge during the Cold War, with most of its elements still in use today. For instance, the modern Russian Army utilizes 122mm, 152mm, and 203mm howitzers, while the 125mm direct-fire self-propelled gun and a 57mm SPAAG are close to entering active service. However, the opportunity to adapt the new system in the late 1940s was lost, as were its predecessors.

In the 1930s, the Red Army pioneered developments in tank and mechanized forces and had the chance to enter World War II with a consistent system of armaments supported by a fully-fledged doctrine. Instead, the steady evolution was disrupted, and the Soviet Army had to rebuild itself in the middle of the war. These specific conditions left their mark on all subsequent developments in this branch of service during and after the war.

Chapter 2

Organization and Order of Battle

Although self-propelled artillery as a distinct class of combat vehicles existed in 1941–42, its structure, organization, or subordination were not clearly defined. The Soviet command at the highest level acknowledged the importance of self-propelled artillery and the need to integrate it into the system of armaments in mid-1941. However, it remained a matter of furious debate within the Soviet military leadership until late 1942.

Nikolay Dmitrievich Yakovlev, head of the Main Artillery Directorate from 1941 and Marshal of Artillery from 1944, remembered in his memoirs:

> Among the senior military leaders of the People's Commissariat of Defense, there were disagreements regarding the determination of the main means necessary to fight with enemy tanks. Artillery commanders insisted on prioritizing artillery, while armored troops commanders insisted on prioritizing tanks. Now, of course, this is already a past dispute. However, it was this dispute that led to the fact that, despite the alarming situation, which further worsened from the mid-1930s, our artillery in infantry divisions remained equipped with 45mm and 76mm guns on horse-drawn carriages.[28]

The beginning of the Great Patriotic War highlighted the importance of mobile means of fire support for infantry and tank units. Again, this sparked another round of discussions. The artillery branch believed that the Red Army needed fast artillery tractors able to tow battalion, regimental, and even divisional artillery. They even proposed to include towed artillery units in the organizational structure of tank and mechanized formations.

Their opponents, in turn, saw a number of problems with this concept. First, unprotected tractors and trucks were vulnerable to enemy fire during the march, as were artillery crews on positions. Another problem was that the Soviet industry simply had no resources or industrial capacity to produce enough tracked tractors for this purpose. Self-propelled artillery, based on serially-produced tank chassis, was therefore a preferable option both economically and tactically.

"Infantry, even without tank support but with self-propelled artillery units, will feel more confident during battles in the depths of enemy defenses than, for example, when they had to wait for horse-drawn artillery or even artillery towed by trucks," wrote Yakovlev.[29]

Another problem "debated for a long time" was the question of subordination. Nikolay Nikolayevich Voronov, Chief of Artillery of the Red Army, believed that self-propelled artillery should be subordinated to the artillery arm and the Main Artillery Directorate

(GAU). This organization was in charge of the development and distribution of all artillery systems and ammunition, including anti-aircraft and anti-tank weapons. In 1941, the Tractor Directorate was also brought under the command of GAU. It is worth noting that this decision was evidently rash and ill conceived, and as a result, the Tractor Directorate was transferred to GAU with no repair facilities for tracked vehicles.

Yakov Nikolaevich Fedorenko, Chief of the Main Automotive-Armored Tank Directorate, in his turn, believed that the tank branch should be in command of self-propelled artillery. His arguments were no weaker. The GABTU supervised all the work on armor and tracked chassis and was in charge of repair and maintenance industry for tracked vehicles. More importantly, self-propelled artillery was technically, doctrinally, and even conceptually closer to the tank forces of the Red Army.

The dispute between the tank and artillery arms of the service was finally resolved on December 7, 1942, when the State Defense Committee (GKO) issued Resolution #2589 to establish a Directorate of Mechanical Traction and Self-Propelled Artillery as part of the Main Artillery Directorate.

An official Order #0972 to establish a new directorate followed on December 21, 1942.[30] This date could be considered as the moment when the Soviet self-propelled artillery was officially born.

The functions of the new Directorate encompassed all questions of development, production, supply, and maintenance of the self-propelled guns, tracked tractors (including the foreign models), and trailers for them. The organization was also tasked with gathering and generalizing the production experience and implementing measures to improve "means of mechanical traction and self-propelled artillery."

Finally, the GABTU was ordered to transfer to the Main Artillery Directorate four repair facilities (#9, #81, #29, and #83) and a 42nd separate repair and maintenance battalion. Therefore, the Self-Propelled Artillery Directorate finally received long-needed repair capabilities.

It is worth noting that while the document covered mostly all important aspects of development, production, and support, the questions of combat employment and doctrine development were not mentioned there at all.

The formation of the new Directorate should have been complete by December 25, 1942, while the departments at the front, military district, and army levels had to be completed by December 30, 1942. The operative command of the new arm was entrusted to Voronov.[31]

The establishment of the new Directorate finally brought some clarity to the previously undefined status of the self-propelled artillery, however the organizational disorder was still in place.

The question of subordination was ultimately decided on April 23, 1943, when Order #0291, "Regarding the transfer of self-propelled artillery under the command of the Tank And Mechanized Troops Of The Red Army" was issued and signed by Stalin. According to this document, the Main Artillery Directorate was ordered to hand over to the GABTU command and control over self-propelled artillery, involving all its elements and functions, such as production, quality control, personnel administration and training. All existing self-propelled artillery units "at the front and in the Reserve of the Supreme High Command (RGK)" were also resubordinated to the Main Automotive-Armored Tank Directorate. In total, 30 regiments were placed under the new command.[32]

Some significant changes were also made in the organizational structure of the staff of the Commander of the Tank and Mechanized Forces, where the posts of Deputy Commander for Artillery, Inspector General for Artillery, Senior Inspector, and his assistant were established.

The Command of Tank and Mechanized Troops, therefore, received full control over the new arm of service, including the responsibilities, resources, and decision-making authorities. The only aspect that remained under the command of the Main Artillery Directorate was the development of artillery weapons and devices, such as optics.[33] This step was more than logical, as it strengthened centralized command, simplified administration, and provided the GABTU with a full picture of self-propelled artillery development, from experimental vehicles to combat deployment.

There is, however, evidence that Stalin was not satisfied with the strengthening of the GABTU. Yakovlev recalled in his memoirs: "In the summer of 1943, the Supreme Commander-in-Chief [Stalin] proposed to put under my command—that is, include in the structure of the Main Artillery Directorate (GAU)—the Main Automotive-Armored Tank Directorate." The official explanation given by the leadership was that the GAU, which was already involved in the development and production of armament and optics for armored vehicles, could take over the production of armor and engines from the GABTU and, thus, combine the development and production of all components under one roof.

While this explanation might seem rational to some degree, it ignored such aspects as training, knowledge, and combat experience that only tank and mechanized forces had. Moreover, the attempt to perform such a major transformation in the middle of the war and right before the summer campaign of 1943 could have caused serious damage to both the artillery and tank arms of the Red Army.

Stalin's proposal created a lot of confusion; Yakovlev remembered that he tried to rationalize the situation and assumed that Stalin was not satisfied with GABTU's performance, but the true reason remained a mystery. Yakovlev rejected the proposal twice; however, Stalin was persistent and offered him to "think again." It was only after the third explicit refusal, when Nikolay Dmitrievich, in his own words, "allowed for some inflexibility after exhausting all arguments," that Stalin finally accepted. Yakovlev wrote, "I simply stated in conclusion that armored fighting vehicles do not belong to our system."[34]

Instability at the highest level of the organizational structure inevitably affected the formation of organizational structures at lower levels. The establishment and development of the first training center for self-propelled artillery can serve as an excellent illustrative example.

The Training Center for Self-Propelled Artillery (UTsSA) was established in accordance with the GKO decree issued on November 24, 1942, and NKO Order #0881 issued on November 12, 1942.[35] This facility was initially formed following Table of Organization and Equipment #08/160. Within a short period from January to late April 1943, the training center underwent a series of changes in its organizational structure in January, February, and April 1943. The initial structure of the center included 12 officers, 18 commanding staff,[36] 17 NCOs and enlisted men, and 44 civilian personnel. In April 1943, the training center was resubordinated to the Tank and Mechanized Forces and reorganized following TO&E #08/179. The new structure included 37 officers, 53 commanding staff, 45 NCOs, and 84 enlisted men; the number of civilian personnel increased to 53. While the initial staff consisted of 47 military and 44 civilian personnel, after the reorganization in April, the staff of the training center increased to 219 military and 53 civilian personnel. The total number of personnel roughly tripled, while the total number of military personnel experienced a five-fold increase.[37]

This organization functioned for almost a year until April 25, 1944, when the training center was reorganized for the fifth time following Table of Organization and Equipment #010/143. The final variant of the organizational structure was used until the end of the Great Patriotic War.[38]

It is worth noting that the changes in the organizational structure were driven by several parallel processes. Firstly, the self-propelled artillery was essentially built from scratch in late 1942–43. Part of the early-stage changes were prompted by the lack of experience at all levels. Secondly, the changes were a response to the evolving requirements of Tank and Mechanized Troops. For example, the emergence of new models of self-propelled artillery and the development of new tactics. Finally, the changes in the later stages were a result of accumulated experience and were initiated to address identified issues and enhance the effectiveness of training and the deployment of self-propelled artillery. The same processes impacted the evolution of the combat unit structure.

On December 27, 1942 following the resolution of the People's Commissariat for Defense #2662cc and Stalin's directives, the Training Center for Self-Propelled Artillery was ordered to establish 30 self-propelled artillery regiments of the Reserve of the Supreme High Command.

Taking into account the limited resources available for this scope of work, as well as the tight schedule and lack of experience, the command of the Training Center for Self-Propelled Artillery had to work with "a high degree of responsiveness and flexibility" and adapt to changing requirements.[39]

The first draft Table of Organization and Equipment for Guards Shock Self-Propelled Artillery Regiment #08/158 consisted of 3 battalions (divisions), one equipped with SU-122 (SU-35) and the other two with SU-76 (SU-12) SPGs. Each division consisted of 3 batteries, and each battery had 4 self-propelled guns.

In total, the self-propelled artillery regiment (SAP) organized following draft TO&E #08/158 had 566 military personnel, 24 SU-76s, and 12 SU-122 SPGs.

However, the next variant of TO&E #08/158 had substantial differences. Now the SAP had only 307 military personnel and 25 self-propelled guns, of which 17 were SU-76s and 8 were SU-122s. In contrast to the draft version with 9 batteries, the vehicles were organized into 6 batteries.

While this variant was approved by the People's Commissariat for Defense on December 27, 1942, it was eventually abandoned. On January 25, 1943, the Training Center for Self-Propelled Artillery began the formation of the units following TO&E #08/191.

The self-propelled artillery regiment organized following TO&E #08/191 consisted of 289 personnel and 21 self-propelled guns, of which 17 were SU-76s and 8 were SU-122s. Compared to both variants of TO&E #08/158, the regiment was smaller, consisting of only 5 batteries, of which two were equipped with SU-76s and three with SU-122 SPGs. Each battery consisted of four self-propelled guns, plus one vehicle attached to a command platoon.[40]

The first self-propelled artillery regiments were the 1433rd and 1434th. These SAPs had a "mixed" organization and were armed with both light and medium self-propelled guns. However, this composition quickly proved itself ineffective. "Mixed" units required a lot of effort to sustain and supply two different types of vehicles armed with two different gun systems.

Between March and May 1943, the Soviet command decided to finally drop "mixed" organization and went through a transition to a new unified system consisting of light, medium, and heavy self-propelled artillery units and four types of tactical units: brigades, regiments, battalions (divisions), and batteries.[41]

In general, this system, with some insignificant changes, was in use until the end of the war. On the contrary, the organizational structure at the unit level followed the changes in tactics or in the system of armament—for example, the retirement of old systems or the introduction of new ones—and was a subject of frequent organizational changes.

During the course of the Great Patriotic War, all types of self-propelled artillery units were subjected to major organizational changes 14 times. By the end of the war, there were 11 TO&Es for regiments, battalions, and batteries, and 2 (in other sources 3) TO&Es for brigades.[42]

The documents of the Training Center for Self-Propelled Artillery state that frequent changes "posed additional challenges in the process of formation, training, and supplying" the units of self-propelled artillery.[43]

While this statement is definitely true, it is worth noting that in the case of self-propelled artillery, the changes were not as frequent and severe as in the case of tank units. Only in the course of the Great Patriotic War, leaving aside the pre-war changes, the Soviet tank brigades used 16 types, tank battalions used 21 types, and tank regiments used another 9 types of TO&Es.[44]

Table 5. Tables of organization and equipment for self-propelled artillery units.[45]

Unit Type	Abbreviation	Number of TO&E
Brigades	SABr	2
Regiments	SAP	9
Division (Battalion)	SADN/OSADN	1
Battery	–	1

The lowest-level tactical unit of self-propelled artillery was a battery. Usually, batteries consisted of 4 self-propelled guns, however, within the structure of the larger units, there also existed batteries that consisted of 2, 3, and even 5 SPGs.

Separate batteries were mostly used to form replacement batteries, temporary units intended to reinforce frontline units that suffered combat losses.

While it was recommended to employ self-propelled artillery *en masse*, in some instances—for example, while acting in the enemy's operative depth—it was allowed to use separate batteries or even separate SPGs with small groups of infantry.[46]

The formation on the next level was a division (battalion).

The self-propelled artillery battalions (SADN) or separate battalions (OSADN) were intended to bolster the anti-tank capabilities of rifle divisions. These units were formed following the order of the General Staff of the Red Army issued on May 30, 1944.

A typical SADN organized following TO&E #04/568 consisted of 152 personnel and 13 self-propelled guns. Each battery consisted of four SU-76 SPGs (12 in total), while the 13th SPG was attached to the commander's platoon and could be used by the battalion commander.[47]

A self-propelled artillery regiment was the most common type of tactical unit and a "building block" for self-propelled artillery in general.

On January 4, 1943, Order #20, titled "On Strengthening the Firepower of Tank and Mechanized Units and Formations of the Red Army," was issued. As per this order, self-propelled artillery regiments were incorporated into the structure of tank and mechanized corps to increase their firepower.[48]

The initial "mixed" regiments, established according to TO&E #08/158, comprised 307 personnel and 25 self-propelled guns. Of these, 17 self-propelled guns were SU-12 (SU-76), while 8 were SU-35 (SU-122) SPGs armed with 122mm howitzers. Each "mixed" regiment featured six batteries, with four self-propelled guns in each. Specifically, four batteries were equipped with SU-76s, while the remaining two were armed with SU-122s. One SU-76 SPG was attached to a command platoon.

Later in 1943, two additional TO&Es emerged. The first one was TO&E #08/191, presenting an enhanced "mixed" self-propelled artillery regiment with five batteries (two armed with SU-76s and three with SU-122s). The total established strength of the self-propelled artillery regiment amounted to 289 personnel and 21 SPGs, including 9 SU-76s and 12 SU-122s.

The second, TO&E #08/218, delineated the structure of a heavy self-propelled artillery regiment, also known as TSAP. Equipped with SU-152 self-propelled guns, each TSAP comprised six batteries, each housing two self-propelled guns. The established personnel strength of the TSAP was 361, with a total of 12 SPGs. In contrast to the "mixed" regiments, TSAPs featured uniform equipment.

In May 1943, a decision was made to abandon the "mixed" structure in order to enhance command and control. Consequently, three new TO&Es for light (TO&E #010/455), medium (TO&E #010/453), and heavy (TO&E #010/454) self-propelled artillery regiments were introduced.

The new SAPs and TSAPs featured standardized equipment. For instance, the medium self-propelled artillery regiment (howitzer) now comprised four batteries, each equipped with four SU-122 self-propelled guns (16 in total), along with 265 personnel and one T-34 tank.

Thus, the reinforcement of the corps was completed with each corps having three self-propelled artillery regiments—light, medium and heavy. However, the introduction of the new models of self-propelled guns and changes tactic prompted another cycle of organizational changes in October 1943.

New TO&Es for light (TO&E #010/484), medium (TO&E #010/483), and heavy (TO&E #010/482) self-propelled artillery regiments were introduced. The changes primarily focused on two aspects: first, each regiment, regardless of the equipment, now had four batteries; second, the number of SPGs in each battery was increased. The heavy SAP had four batteries with 3 SU-152 SPGs, the medium SAP had four batteries with 4 SU-85 SPGs, and the light SAP had four batteries with 5 SU-76 SPGs.

The latest round of organizational changes was initiated in February 1944. Throughout 1944, the new models of self-propelled artillery were introduced into service. The major models in production included ISU-152, ISU-122, SU-85/100 and SU-76 SPGs. In addition, all heavy self-propelled artillery regiments were redesignated as Heavy Guards self-propelled artillery regiments.

It's worth noting that these changes were implemented gradually, following the supply of new equipment. As a result, during 1944–45, the Red Army had a mix of units utilizing both older and newer TO&Es.

The final structure was as follows: Heavy Guards self-propelled artillery regiments were organized according to TO&E #010/461. Each regiment consisted of 21 SPGs in four batteries, with five ISU-152 SPGs in each battery, plus one ISU-152 in a command platoon.

Medium SAPs followed TO&E #010/462, resembling the structure of the TSAP armed with ISU-152. The regiment featured 21 SPGs in four batteries, with five SU-85 or SU-100 SPGs in each battery.

Finally, light SAPs retained the older TO&E #010/484. These units also comprised 21 SPGs in four batteries, with five SPGs in each battery and one SPG attached to a command platoon.[49]

The highest-level self-propelled artillery formations were brigades. There were three types of them: light brigades armed with SU-76, heavy guard brigades armed with ISU-122 or ISU-152 SPGs, and medium brigades armed with SU-100s.

On February 5, 1944, in accordance with a directive issued by the General Staff of the Red Army, the self-propelled artillery brigades were established following TO&E #010/508.

A light brigade included 1122 officers and enlisted men, 60 SU-76 SPGs, 5 T-70 or T-80 light tanks, and 3 armored personnel carriers or BA-64 armored cars.[50] Self-propelled artillery was organized into three battalions (divisions), plus a submachine gunner battalion of 298 men. Each artillery battalion had four batteries of five SU-76 SPGs (20 SPGs in total), plus a support platoon.

Heavy and medium self-propelled artillery brigades were established later, between December 1944 and January 1945. A Guards heavy self-propelled artillery brigade comprised of 1804 officers and enlisted men, 65 ISU-152 or ISU-122 SPGs, and 3 light SU-76 SPGs. Each heavy brigade included three heavy self-propelled regiments organized according to TO&E # 010/461. Each regiment consisted of 4 batteries of 5 SPGs (20 SPGs in total per regiment), a submachine gunner company of 110, a repair and maintenance company of 82, and a command platoon of 43 with one self-propelled gun. Two more heavy SPGs were subordinated to a brigade-level command company, while three SU-76 SPGs were subordinated to a reconnaissance company.

Finally, a medium self-propelled artillery brigade had an established strength of 1492 officers and enlisted men, 65 SU-100s, and 3 light SU-76 SPGs. The brigade was broken down into three regiments; each of them was organized into four batteries of five SU-100 SPGs and a submachine gunner company of 60. On top of that, each command platoon at a regimental level had one SU-100 SPG. As in the case of a heavy brigade, two SU-100 and three SU-76 SPGs were attached to command and reconnaissance companies at brigade level.[51]

Self-propelled artillery brigades were intended for use as force multipliers for combined arms and tank armies, i.e., at the operative level of command.[52]

Between late 1942 and 1945, self-propelled artillery not only was reestablished but also found its place within the system of the Ground Forces of the Red Army and proved itself as an invaluable asset for infantry and tank troops. However, the self-propelled artillery and its organizational structure had to evolve at a rapid pace during the war; for this reason, mistakes and frequent changes were inevitable.

The "Report on the Development of the Tank and Mechanized Forces of the Red Army During the Great Patriotic War" highlights seven phases of the development of the organizational structure. The first stage was the establishment of the "mixed" self-propelled artillery regiments armed with SU-12 and SU-35 SPGs in late 1942. The second stage began in April 1943, when the self-propelled artillery was resubordinated to Tank and Mechanized forces and the organizational structure changed to a unified system with three types of regiments: light, medium, and heavy. The third stage began on August 28, 1943, when the self-propelled artillery regiments were incorporated into the structure of tank and mechanized corps as force multipliers. The next, fourth stage, was initiated in October 1943. This was a major change in the organizational structure aimed at bolstering the firepower of the self-propelled artillery batteries. The structure of a self-propelled artillery regiment with six batteries was abandoned, while the new structure included only four batteries with more SPGs. The number of vehicles in a heavy battery increased from 2 to 3, while the number of light SU-76 SPGs increased from 4 to 5. At the fifth stage, after the Directive of the General Staff of the Red Army was issued on February 28, 1944, heavy and medium self-propelled regiments were reorganized to incorporate new materiel: ISU-152 heavy SPGs and SU-85/100 medium SPGs. At the sixth stage, the TO&E for self-propelled artillery battalions (divisions) was introduced on May 30, 1944. The final, seventh stage included the introduction of TO&Es for light, medium, and heavy self-propelled artillery brigades. These changes were implemented between late 1944 and early 1945.[53]

As can be clearly seen, in just two and a half years, self-propelled artillery evolved from a tactical-level to an operational- and even operational-strategic-level asset. Self-propelled artillery units became an integral part of Soviet tank and mechanized troops and were used as force multipliers with units from rifle divisions to combined arms and tank armies, thanks to their unique combination of combat capabilities.

The number of active self-propelled artillery units on June 1, 1945, speaks for itself. By this time, the Red Army had 13 self-propelled artillery brigades, 249 self-propelled artillery regiments, and 70 separate self-propelled artillery battalions.[54]

Another important development was the implementation of self-propelled artillery battalions into the structure of rifle divisions, and in fact, it was a return to the pre-war concept of artillery mechanization pioneered by Mikhail Tukhachevsky, Konstantin Kalinovsky, Vladimir Triandafillov, and others.

The intention was to replace the towed artillery of rifle divisions with self-propelled artillery battalions armed with SU-76 SPGs and, therefore, to provide rifle divisions with organic mobile means of fire support.

While this process was not finished before the end of the Great Patriotic War, it laid the foundation for further changes. For example, in 1947, the so-called "tankosamokhodnye" regiments were integrated into the structure of the Soviet rifle divisions and brigades. These mixed mechanized regiments included T-34/85 medium tanks and SU-100 or SU-76 SPGs and were meant to provide rifle divisions with a versatile combat capability combining firepower, mobility, and protection.

Chapter 3

Training: Sweat Saves Blood

There was no specialized training for self-propelled artillery in the Red Army until late 1942. The first training facility for self-propelled artillery, the Training Center for Self-Propelled Artillery (UTsSA), was established in November 1942. While the training system as a whole transformed several times over the course of the Great Patriotic War, the UTsSA remained the primary training center for self-propelled artillery.

In April 1943, the UTsSA, with "all personnel, training materiel, warehouses, and self-propelled artillery regiments in formation," was transferred under the command of the GABTU, in accordance with Order #0291. Under the same directive, the 18th, 19th, and 21st Training Self-Propelled Artillery Regiments, the 2nd Kiev and 2nd Rostov Self-Propelled Artillery Schools with all personnel and training infrastructure, and personnel from the Artillery Unit Formation Directorate and the 3rd Department of Combat Training Directorate involved in combat training for self-propelled artillery were also moved under the command of GABTU.[54]

Between January 1, 1943 and January 1, 1944 the training system of the self-propelled artillery had the following structure. Initially, the Training Center for Self-Propelled Artillery was responsible for forming, equipping, and training self-propelled artillery regiments and replacement batteries and sending them to the front. Over time, the responsibilities were extended. According to TO&E #08/179, the UTsSA included five so-called "groups of formation." One group was responsible for training and forming heavy self-propelled units, while two groups were assigned to medium self-propelled units, and another two groups focused on light self-propelled units. In May 1944, the organizational structure of the UTsSA was changed. According to new TO&E #08/143, the number of groups of formation was reduced to four. This structure was used until the end of the war.

The 18th, 19th, and 21st Training Self-Propelled Artillery Regiments were responsible for training drivers for SU-76, SU-122, and SU-152 SPGs. All three regiments were stationed in cities where the self-propelled guns were manufactured: the 18th Regiment in Kirov, near the Plant #38 manufacturing SU-76 SPGs; the 19th Regiment in Sverdlovsk (now Yekaterinburg), where the UZTM plant was situated; and the 21st Regiment in Chelyabinsk, where the Chelyabinsk Kirovsky Zavod (ChKZ) was located.

The 15th Reserve Self-Propelled Artillery Regiment was transferred under the command of the UTsSA in February 1943. This regiment was in charge of training junior specialists for self-propelled artillery.[55]

The 1st Special Purpose Self-Propelled Artillery Regiment (so called "literny" regiment) was established on December 10, 1943. This unit was attached to the 3rd group of formation of the UTsSA and was tasked to train crews for foreign self-propelled artillery, including the M15, M17, T48 and M10.

In the autumn of 1943, the demand for self-propelled artillery specialists had outgrown the capabilities of the existing training system. In order to increase training capabilities, the 1st and 5th Training Tank Brigades in Gorky (now Nizhny Novgorod) and Sverdlovsk (now Yekaterinburg) were reorganized and became the 1st and 5th Training Self-Propelled Artillery Brigades. In addition, the 21st Training Self-Propelled Artillery Regiment in Chelyabinsk was also reorganized and became the 7th Training Self-Propelled Artillery Brigade. The primary task for these brigades was forming the replacement batteries.

In addition, following the order of the Commander of the Tank and Mechanized Forces the UTsSA was authorized to temporarily include in its structure three self-propelled artillery regiments: 1456th SAP attached to the 4th group of formation, 1460th with the 3th group of formation and 1457th with the 1st group of formation.[56]

The officer training was organized at the 2nd Kiev and 2nd Rostov Self-Propelled Artillery Schools. In July 1941, the 2nd Kiev Artillery School was relocated to the village of Razboishchina, Saratov Oblast. On January 16, 1943, the institution was reorganized and officially redesignated as the 2nd Kiev Self-Propelled Artillery School. This training center became the primary institution for training officers for self-propelled artillery, with a capacity of 1,450 cadets. The 2nd Rostov Self-Propelled Artillery School was established on February 13, 1943, and was able to train 1,600 cadets. However, the problem of inadequate training capacity had also resurfaced in the case of officer schools. During the period between the autumn of 1943 and the spring of 1944, the system was supplemented with eight more officer training schools.

Essentially, the training system of self-propelled artillery, together with the whole system of Tank and Mechanized Forces, was going through multiple changes, adapting to the changing demands and requirements. As a result, the stable and structured, while not completely flawless, version of the training system emerged only at the beginning of 1945.

On January 1, 1945, this version included an UTsSA with four groups of formation: three training brigades (1st, 5th, and 7th); three training regiments (32nd, 37th, and 42nd); and two reserve regiments (15th and 36th). The 14th Self-Propelled Artillery Brigade was attached to UTsSA for training replacement batteries and regiments.

The officer training system included nine self-propelled artillery schools: 2nd Kiev, 3rd Gorky, Solikamsk, Buisk, 2nd Rostov, Kotlas, Chkalov school in Kharkov, Syzran, and 2nd Ulyanovsk.

On top of that, four tank training camps were also involved in training: the Moscow (Naro-Fominsk before April 25, 1944), Tula, Belorussky, and Ukrainsky camps. The Naro-Fominsk Tank Training Camp was the main training facility for the foreign armored combat vehicles received through Lend-Lease. Of the 13,068 tanks and self-propelled guns received by the Moscow tank training camp, 6220 were of foreign origin.[57]

Outside the training system, the frontline troops also had their own mixed tank-self-propelled artillery regiments aimed at rapid crew training for replenishment of the units that suffered severe combat losses.

Establishing and operating a training system of that size and complexity is a challenging task in itself. In the particular case of the Soviet Tank and Mechanized Forces, the entire system for self-propelled artillery had to be built practically from scratch under conditions of war.

Although the training system was created and functioned at a remarkable pace, a number of factors affected it over the course of the war. The first major problem was the availability of qualified manpower. First, in early 1943, there were no instructors with specialized experience and knowledge able to train personnel for self-propelled artillery units. Furthermore, the Command of the UTsSA struggled to find and attract qualified officers for leading positions, especially for the Combat Training Department. The reason for that was "low salary rates."[58]

The other side of the same problem was the inadequate quality of the available manpower. According to the UTsSA and tank training camp reports, "privates and sergeants arriving [to the UTsSA] from the front, reserve, and training units for formation have inadequate training, and this requires training them individually before proceeding to the unit-level training programs."[59]

The report highlights that the problem was especially severe in 1943, which was understandable considering the significant growth of the Tank and Mechanized Forces and self-propelled artillery during this period. According to the same documents, this problem was typical for any military profession: commanders, tank and truck drivers, gunners, sappers, submachine gunners, and others[60]

Another personnel-related problem was the scarcity of time. Sometimes the personnel arrived at the training camps for only 2–3 days before the newly formed unit departed for the front. As a result, the instructors were unable to familiarize themselves with the soldiers, evaluate their level of training, and fill the gaps in their skills and knowledge.

The next problematic area was the low level of tactical and special training among the officers arriving at the self-propelled artillery units under formation. This problem was flagged up among Battery Commanders, their Deputy Technical Officers, and self-propelled gun commanders. According to the UTsSA report, this problem "severely complicated the training of crews, batteries, and units and required additional efforts to rectify this situation." Usually, the command of the training centers had to organize additional training exercises and classes, which was not always easy or possible at all.[61]

The next challenge was the lack of infrastructure. Already in 1942, the UTsSA faced a shortage of housing to accommodate its personnel and provide sufficient space for training classes. As a result, the UTsSA had to find additional manpower and resources for construction works.[62] For example, according to the report of the Moscow Tank Training Camp, due to the shortage of housing, a significant part of the units coming from the front had to interrupt their training and build dugouts, which sometimes took 10 to 15 days. This, in turn, negatively affected their combat training.[63] Additionally, the sites where the groups of formations were located were surrounded by land pieces owned by state organizations or collective farms ("kolkhoz"). Whenever possible, the groups of formations used this land for training purposes. Since collective farms needed their land for agricultural needs in the summer, "the number of training grounds and firing ranges steeply reduced, making tactical regimental or brigade-level training difficult or impossible."[64]

This situation might seem impossible within the overcentralized Stalinist system with its meticulous planning; however, it was the direct consequence of it. Both the military and civil systems had their own plans and goals to achieve. Failing to fulfill the plan, especially in times of war, was a serious offense and could lead to serious consequences in different forms, from fines or reprimands to being charged with negligence or sabotage. For that reason, the leadership of the collective farms was more concerned with the fulfillment of their own plans, even if it led to serious changes in the military training plans.

Probably the most severe and paradoxical problem that the UTsSA faced was that the self-propelled artillery training center did not have its own artillery range. The UTsSA was allowed to use the artillery proving grounds of the People's Commissariat of Ammunition (NKB) at Sofrino, near Moscow. However, the UTsSA command had to harmonize their training schedule with the NKB, and, according to the UTsSA leadership, "the NKB proving grounds were not able to ensure training at all times and, due to its size, it did not allow large-scale tactical training, especially bilateral or combined arms training involving fire and maneuver."[65]

The combination of geographical scale and weak means of communication posed additional difficulties for the training system. The UTsSA alone was deployed at several sites in Mytishi, Pushkin, and Zagorsk districts near Moscow. On top of that, the units and subunits were often redeployed, causing additional troubles. For example, from January 1944 to July 1944, the group of formation for heavy self-propelled artillery was deployed in Naro-Fominsk, i.e., 100km away from the UTsSA command. This was exacerbated by the absence of the communications unit within the structure of the Staff of the UTsSA.[66]

Another obstacle was inadequate training programs. "The combat training programs for units and formations received from the Combat Training Directorate were developed in the beginning of 1943, when self-propelled artillery was not widely used at the fronts of the Great Patriotic War, and hence were not built on combat experience," says the UTsSA report.[67] As a consequence, these training programs were already obsolete by the end of the same year. Additionally, the old programs were developed for a full course lasting 10 to 15 days. However, the UTsSA did not have approved programs for longer periods of formation. In these cases, instructors had to repeat the material covered earlier or take a risk and use unapproved courses.

Since the Combat Training Directorate was not able to deliver programs and manuals for self-propelled artillery in time, the UTsSA had to develop and use their own training programs. While it was somewhat more effective as the UTsSA officers better understood the real needs of the self-propelled artillery units, the training center had to reallocate specialists to this task, which "affected the efficiency of training."[68]

The insufficient number or, in some instances, complete absence of training programs, combat manuals, pamphlets, periodical magazines, and even visual guides was a further setback that plagued the training process over the course of the war. The UTsSA highlighted this problem, stressing that "the inadequate number of visual aids, periodical literature, and the absence of new combined arms combat manuals and recommendations negatively affected the training of the personnel arriving from the reserve and training units."

The shortage of service manuals for foreign equipment was a unique aspect of this problem. The problem became extremely serious when the training center was ordered to train personnel for US-made self-propelled guns.[69] In addition, the units attached to UTsSA did not have a fleet of permanently deployed training vehicles. For that reason, the training center had difficulties delivering practical training for drivers in replacement units, mechanics, and armored vehicle technicians."

The most significant disadvantage, according to the UTsSA leadership, was the absence of a combat manual for self-propelled artillery. The questions of doctrine and tactical employment will be examined in detail in the next chapter; however, the UTsSA mentioned that this problem 'significantly complicated the tactical training, especially in questions of combat employment, combined arms warfare, combat formations, and command and control." When the officers arrived at the Training Center for Self-Propelled Artillery, the instructors had no clear guidelines on how and what to teach. The officers assigned to self-propelled artillery, in turn, had mixed experience and training; part of them were tankers and the other part were artillerymen. This lack of guidance proved to be a painful experience, resulting in less effective and more complicated training.[70]

Finally, a chronic planning problem persisted throughout the war. At the highest level, according to the UTsSA report, the Directorate for Formation and Manning of the Tank and Mechanized Forces (UFiU) did not provide the training system with any schedules or plans. In some instances, the UTsSA received an order to form and transfer a unit via phone call.[71]

This ad hoc system produced a lot of confusion, and the terms for formation were rarely met. Sometimes the self-propelled artillery units that finished their training in time had to wait for the materiel to arrive. In other, more common cases, the organizational and administrative chaos resulted in reduced time for training.

This, in combination with demands set up by the front-level command, resulted in severely reduced time for collective training. Already tight schedules provided by the Combat Training Directorate were cut twice or more. In these cases, the units had only 3–5 days of collective combat training after formation and receiving the materiel. According to the report of the Moscow Tank Training Center, "the collective training began with receiving armored vehicles from the factory and ended with loading on a train."[72]

The Tula Tank Training Camp reported that "90% of personnel were late for formation from 5 to 20 days. There were cases when a unit departed to the front and joined the fight while its personnel continued arriving at the camp for formation."[73] In extreme scenarios, organizational disorder led to last-minute changes. For example, the 22nd Nevelskaya Self-Propelled Artillery Brigade received a new TO&E during loading on the train and was hastily reorganized the same day.[74]

The systemic organizational disorder disturbed the operations of the training system as a whole, resulting in ineffective resource management and, eventually, lowering the efficiency and quality of training. Nevertheless, while the training system for Tank and Mechanized

Forces had its share of shortcomings, it effectively facilitated the rapid and extensive training, formation, and organization of self-propelled artillery units on short notice.

Table 6. Number of officers trained for self-propelled artillery in 1944.[75]

	From 1 Jan 1943 to 1 Jan 1944
Tank schools and courses	3188
Military academies	18,074
Officer Professional Enhancement Courses	1982
Total	23,244

Table 7. Number of crews trained for self-propelled artillery in 1943–44[76]

	1943	1944
Heavy SPGs	661	2331
Medium SPGs	1449	2450
Light SPGs	2245	7690
SU-57 (T48)	–	561
M15	–	116
M17	–	1190
Total	4355	14,338

Table 8. Number of self-propelled units formed and replenished in 1943–45, according to UFiU data.[77]

	Formed	Replenished
Heavy self-propelled artillery regiments	30	80
Medium self-propelled artillery regiments	50	71
Light self-propelled artillery regiments	69	91
Separate self-propelled artillery brigades	11	3
M15 and M17 replacement batteries	78	0

Following Order #0436 issued on October 16, 1943, and signed by Stalin, a period of six months was established for training personnel in all training units of the Red Army, including tank and self-propelled regiments.[78] Additionally, the official period of 10 to 15 days for collective combat training was established. While these terms were sound, due to the dynamics of warfare, front-level command requirements, and other factors mentioned above, they were rarely met.

The training process itself was multilayered and complex. The lowest tier was individual junior specialist training. This part of the training system was essential to prepare specialists for self-propelled artillery as well as train or retrain specialists in general military professions, such as drivers or sappers. For example, the 15th Reserve Regiment of the UTsSA was responsible for training 53 different military professions, from technicians and mechanics to truck drivers, radio-masters, and even cooks.[79] Sometimes the soldiers and officers assigned to self-propelled artillery already had previous training. In this case, they needed additional courses to get accustomed to the specifics of tank and self-propelled artillery.

Another form of individual training was short-term classes for specialists, such as ammunition supply officers, signal officers, personnel officers, reconnaissance officers, and others. The main goal of these classes was advanced professional training, involving familiarization with combat manuals, orders, and recommendations, and practical drills with equipment.

The next level was collective training. The most common drills were battery-level drills, which, whenever possible, included field exercises with equipment. At times, exercises included live-fire drills, while other times, they focused solely on maneuvering.

Brigade- and regimental-level field exercises were held less often for two reasons. First, self-propelled artillery brigades were added to the organizational structure only in February 1944; second, large-scale exercises were harder to organize and required more time and resources.

Collective training consisted of three stages. The first stage commenced before the unit received its combat equipment; its purpose was to prepare unit officers to act as instructors for their personnel, establish a staff, and prepare a training plan.

The second stage commenced after receiving combat materiel and focused on collective combat training for crews, platoons, companies, and ultimately, the entire unit or formation. The primary objective of this stage was to provide the unit with collective combat training depending on time available, and to prepare the unit for various combat scenarios, including defensive and offensive actions, ambushes, and deployments, among others. Additionally, units were taught to act in different weather conditions and on different terrain.

The third stage commenced after the conclusion of collective combat training. It served as an opportunity for providing advanced training and revisiting topics in case supervising officers found the training on specific subjects to be unsatisfactory. Importantly, this stage did not include live-fire exercises or driving exercises.[80]

It is worth noting that throughout the war, the issue of the economy of resources was one of the most pressing. The training system faced shortages in various resources, such as motor hours and live ammunition for tanks and self-propelled artillery.

For example, the leadership of the Tula Tank Training Camp in the post-war report offered to set a limit of 10 motor hours and 6 projectiles per crew.[81] While the officers believed the limits to be adequate based on their own experiences, there were instances during the war when even these numbers were not met.

The officers received their education at the specialized officer training schools. Additionally, they attended courses and so-called demonstration drills. These drills involved live exercises in which a unit conducted collective training, while a group of officers and their staffs from other units observed the drills under the supervision of an officer-instructor from the training center. The instructors assisted them in evaluating the positive and negative sides of the unit's actions. This assessment encompassed various elements, including fire control, terrain analysis, command and control during combat, and armored maneuvers on the battlefield. Whenever feasible, bilateral drills were arranged.

Another vital aspect of the training system involved leveraging the experiences gained by self-propelled artillery units on the front lines. These experiences were collected, analyzed, and frequently incorporated into training. The use of real-life examples from small units trained and organized at specific training centers, along with insights from officers who had participated in combat, had a profoundly positive impact on cadets.[82]

Table 9. Number of training exercises per type conducted at the Training Center for Self-Propelled Artillery, November 1942–May 1945.[83]

Live-fire tactical exercises, brigade level	4
Live-fire tactical exercises, regimental level	262
Live-fire tactical exercises, replacement batteries	1560
Programs for collective training (6 days to 1 year)	34
Short-term (1–3 days) tactical drills without live fire	272
Short-term classes, including for experience exchange	24
Letters for experience exchange sent to frontline self-propelled artillery units	364
Reports received from the frontline self-propelled artillery units and studied	116
Recommendations developed by the Department of Combat Training	14
Tactical recommendations developed for brigade and regimental-lever exercises	15

Table 10. Number of training exercises per type conducted at Tula Tank Training Center, August 10, 1941–May 1945.[84]

Unit and formation-level exercises	743
Demonstration drills	304
Instructional-methodological training	203
Combat experience exchange conferences	14
Short-term classes for specialists	17
Reports on combat experience exchange	31
Plans and abstracts for officers	64
Live-fire exercises	640

The typical 10-day training course for self-propelled artillery brigade armed with SU-100 SPGs included the following topics:

- Preparing SPGs for combat;
- March of the self-propelled artillery battery;
- Actions of a single SU-100 SPG from ambush with a live-fire exercise;
- Actions of the SU-100 SPG battery repelling enemy tank counterattacks with a live-fire exercise;
- Actions of the SU-100 SPG battery in repelling enemy tank counterattacks, including tactical drills with radio communications;
- Actions of the SU-100 regiment supporting the offensive actions of the tank corps;
- March of the self-propelled artillery regiment;[85]

While these topics may seem basic, they surely laid the groundwork for the future training and ensured at least the minimum required knowledge and skills.

Army training was the final and arguably the most crucial part of the training cycle, serving two vital functions.

Firstly, the system enabled troops to receive advanced training based on lessons learned. Frontline units shared their accumulated combat experience with reinforcements and even provided tailored advanced training based on their own experiences, especially when it came to specific equipment or environments. Courses in the self-propelled artillery units were organized whenever possible, primarily during lulls in the fighting.

Training programs were typically planned by unit staff according to the available recommendations and manuals. Then, the programs had to be approved at a higher level of command. For example, programs for the 12th Self-Propelled Artillery Brigade (SABr) were approved by the Commander of the Tank and Mechanized Forces of the 69th Army.[86]

The duration and content of the training courses varied depending on many factors, such as available time, the operational environment, and not in the least, the skills and organizational abilities of the officers in a certain unit. Usually, training periods were planned for 8–10 days with a total of 80–100 hours of training per period, averaging 8–10 hours per day. However, it should be taken into account that soldiers had to fulfill their duties and could be reallocated to perform other tasks, which left them with less time for training.

The wartime documents of the 12th SABr offer a vivid example. Initially, the program for the self-propelled artillery battalions was planned for a total of 98–102 hours and included 10 different disciplines. However, during the training period, crews were busy entrenching their SU-76 SPGs, building dugouts, moving materiel and equipment to other positions, and conducting reconnaissance missions.

In order to address this issue, the command of the 12th SABr found a balanced solution. They omitted less necessary disciplines, such as chemical training, marching drills, and knowledge of military regulations, while focusing on specialized disciplines necessary for combat, including tactics, gunnery practice, topography, equipment maintenance, and others.

As a result, the brigade only completed 71.9% of the course. The most important disciplines and topics were covered to a satisfactory level among the SPG crews. However, other elements of the brigade, such as the submachine gunner company, command, maintenance, anti-aircraft machine gun companies, and medical platoon, received less training and only covered 25.6% of the program.[87]

Another function was to complement the training system by providing additional guidance to inexperienced troops arriving from training camps and schools. This was necessary because the quality of initial training in the rear was often deemed inadequate. It would not be an exaggeration to say that the army training system bore the major burden of addressing flaws within the entire system.

It was not uncommon for frontline units to almost completely retrain the reinforcements with the basics before moving on to advanced-level training. The documentation of the 377th Guard Heavy Self-Propelled Artillery Regiment (TSAP) provides us with a striking example of the low-quality training that negatively affected the combat efficiency of the whole unit.

This regiment was armed with SU-152 SPGs, and in April 1945, its command organized tactical training with live-fire drills. After the drills, the instructors reported that the crews, who arrived with replacement batteries from the rear, were poorly trained in gun laying, working with gun optics, range finding, gun loading, and other essential skills. Consequently, they were able to open fire 10–15 minutes (!) after the order to open fire was given.

This particular instance also highlights the difference between the technical characteristics on paper and the real-world capabilities of the equipment. While the rate of fire for the SU-152 SPG is often cited as 1–2 rounds per minute, the real-world rate of fire was significantly lower and strongly depended on the level of crew training.

Another problem was the SPG commanders' training. Instructors noted that they were unable to give orders in accordance with regulations, had inadequate target designation and fire control skills, and were not able to adjust fire or work with radio stations.[88]

Overall, the Soviet training system allowed for the provision of at least minimally trained crews to fill the ranks of the growing tank and mechanized forces. On the other hand, the system was unnecessarily complex, implemented hastily, and had numerous drawbacks. Quantity of trained troops and formed units often dominated over the quality of training throughout the war.

In November 1944, the poor quality of training for tank and SPGs crews drew sharp criticism from Marshal Georgy Zhukov and Commander of the Tank and Mechanized Forces of the Red Army, Yakov Fedorenko. A directive titled "On Tank and Mechanized troops training in 1944" stated that "Marshal of the Soviet Union Georgy Zhukov personally established that individual combat training of infantry soldiers small infantry units, self-propelled artillery, tanks and artillery is very poor and lags behind modern wartime requirements in many ways. As a result, privates and sergeant personnel are not trained in the elementary rules of combat, and we are incurring needless losses."[89]

Additionally, Marshal Fedorenko and Colonel-General Biriukov provide an extensive list of drawbacks identified in tank and mechanized troops, such as the commanders' poor evaluation of terrain advantages in the context of their combat tasks, inadequate navigation skills—especially when engaged in enemy territory or under challenging conditions like adverse weather, limited visibility, or nighttime operations.[90]

When discussing platoon and company commanders, they emphasized their insufficient tactical training and lack of knowledge regarding the technical and tactical capabilities of various branches of service, such as infantry or artillery. This lack of understanding of their respective methods of warfare often resulted in inadequate task assignments, leading to inefficient coordination on the battlefield. Fedorenko and Biriukov discussed the quality of training for tank and self-propelled artillery commanders. They stated that these commanders often lose communication, have poor situational awareness, and consequently fail to show initiative. Instead, they remain inactive, stagnating, and waiting for instructions from infantry commanders or their senior officers. Additionally, tank commanders and self-propelled artillery commanders lack the knowledge of how to utilize terrain folds for movement and halts during firing.[91] The document suggested a reason for this state of affairs: "all this happens because those who provide training are inadequately trained themselves; they have a limited understanding of modern combat and do not study it sufficiently."[92]

Almost immediately, on November 22, 1944, Fedorenko and Biriukov issued another document titled "Directive of the Commander of the Tank and Mechanized Forces for Improving Training for Tank and Self-Propelled Artillery Units." Both directives called for immediate action and ordered the implementation of improvements in every aspect of training. This included basic tactical training and knowledge of field manuals, with a strong emphasis on practical training, such as night driving, maneuvering and gunnery.[93] While the problem was acknowledged and understood at the highest level of command, implementing improvements on such a large scale proved challenging, especially given objective limitations such as a lack of resources and time. The problem of training quality was not fully resolved until the end of the Great Patriotic War.

Chapter 4

Tactics and Doctrine

As mentioned in Chapter 2, self-propelled artillery became an independent arm of service only in late 1942. Prior to that, the Red Army had limited experience with self-propelled artillery, and as a result, there was no established doctrine or formalized recommendations regarding its combat employment or tactics. Furthermore, the vast majority of self-propelled artillery employed by the Red Army in small conflicts during the 1930s consisted of indirect fire artillery, whereas most of the vehicles used during the Great Patriotic War were direct fire self-propelled guns. Pre-war knowledge was therefore only partially relevant and useful.

On January 5, 1943, the Main Department of the Commander of Artillery of the Red Army issued "Temporary Regulations for the Combat Employment of Self-Propelled Artillery."[94] The importance of the document was tremendous, as it was the first doctrinal document since the 1930s and the Deep Battle doctrine that formalized operational procedures, tactics, and techniques. The "Temporary Regulations" provided general instructions on the employment of self-propelled artillery units in offense and defense; self-propelled artillery supporting independent tank formations; and even the place of the self-propelled units during a march with infantry or tank units. Although there were some shortcomings and oversights, it laid the necessary groundwork for the initial combat employment of self-propelled artillery.

On the other hand, the document was very brief and omitted many fundamentals. For example, it did not include a clear definition of self-propelled artillery, offering only an ambiguous description.

According to it, self-propelled artillery combined high mobility, annihilating firepower, and armored protection, which ensured a high level of readiness, speed, and the ability to surprise the enemy on the battlefield. Essentially, it shared the same combination of three key characteristics of tanks, or, broadly, armored vehicles, known as the "Iron Triangle."

The lack of a clearly defined distinction between self-propelled artillery and tanks could lead to confusion. To address this issue, the authors added a special paragraph for clarification, warning that despite any external resemblance between self-propelled artillery and tanks, as well as some similarities in their employment, these two modern weapons should never be equated in their tactical use under any circumstances.[95]

There were also no specific instructions regarding the employment of various types of self-propelled artillery—light, medium, and heavy—as well as its specific variants, such as indirect fire, anti-tank, or anti-aircraft guns. Some instructions on specialized subjects were broadly formulated, lacking clarity and details. For instance, there was a section that addressed the question of ammunition supply on the battlefield. The regulations recommended having trailers for each self-propelled gun—wheeled during the summer and sleighs during the winter—equipped with prepared ammunition crates.[96] However, the "Temporary Regulations" did not offer any further instructions regarding the supply of self-propelled artillery on the battlefield using unarmored trailers or sledges. It provided no explanation of the exact supply procedures or details about the place of this equipment in the organizational structure. The only exception was a paragraph stating that "all tasks, including the order of ammunition supply, should be determined in the field." In effect, the regulations placed the responsibility for decision-making on tactical commanders, who, at the time of the regulations' issuance, had no prior experience with self-propelled artillery.[97] Finally, the concept of using SPGs for towing ammunition trailers contradicted the principle of mobility, one of the basic principles of self-propelled artillery.

The document also offered only brief recommendations on tactics and combat employment. According to the regulations, self-propelled artillery was meant to accompany tank, infantry, or cavalry units, or to act as maneuver artillery reserves for the tank or combined arms commander.

The main tactical role outlined for self-propelled artillery was direct fire support. When acting as a mobile reserve, self-propelled artillery should have been employed against enemy targets at close range, along the main axis of advance, and during decisive moments of infantry, tank, or cavalry attack.[98]

In offensive operations, self-propelled artillery units were meant to attack enemy positions, destroying targets that prevented allied infantry (tanks, cavalry) from advancing. Ironically, this was exactly the same role envisioned for direct infantry support tanks, or NPP tanks, in the Deep Battle doctrine developed in the early 1930s.[99] In defense, self-propelled artillery's task was to annihilate the enemy forces that managed to break through the defensive lines.

Along with the fundamental tactical guidelines, the regulations introduced a few innovative concepts, although these concepts had not yet been battle-proven.

The first one was a roving gun tactic, known today as shoot-and-scoot.[100] The regulations recommended exploiting the high mobility of the SPGs to take favorable positions, conduct a fire mission, and quickly relocate vehicles to another position to avoid counter-fire or detection by the enemy.

The second innovation was centralized and decentralized employment of a self-propelled artillery regiment.

The former implied using a regiment *en masse*, for example, on a main axis of attack, while the latter allowed for breaking up the regiment and attaching batteries, platoons, and sometimes even single vehicles to infantry or tank units for support.[101]

To allow "quick transition from decentralized to centralized employment and vice-versa," the regulations repeatedly stressed the importance of keeping stable and reliable communications.[102] The self-propelled artillery regiment commander had to ensure communications with the higher-level staff of the tank or combined arms formation. Equally important was keeping stable communication lines with battalions, batteries, and down to a single self-propelled gun crew level.

Theoretically, the concept was sound, however it required a high level of personnel training and equipping the self-propelled artillery units with high-quality communication equipment. Achieving both conditions posed significant challenges and eventually, the decentralized method of employment proved itself flawed and ineffective, although it was used regularly.

Summing up, it can be said that the "Temporary Regulations" was a raw document developed without sufficient practical experience and implemented under time constraints. The first two self-propelled artillery regiments, the 1433rd and 1434th, were sent to the Volkhov front in late January 1943, less than a month after the "Temporary Regulations" was issued, leaving no time for adjustments.

The first battles predictably exposed the imperfections in the doctrine and "Temporary Regulations" in particular and inadequate training of SPG commanders and crews. While the problems were reported and discussed, the system was going through organizational changes and the deficiencies remained largely unresolved.

The Battle of Kursk provided more valuable practical experience and again raised concerns regarding doctrine and training. On July 25, 1943, the Commander of the Bryansk Front mentioned in his order that "during the last days of combat there were cases when self-propelled artillery regiments were employed incorrectly, as direct artillery support tanks." More importantly, self-propelled artillery regiments were tasked with attacking enemy defenses independently, resulting in "unnecessary losses in men and materiel."[103]

Some initial difficulties with the tactical use of self-propelled artillery were understandable, given that the Battle of Kursk marked the first large-scale deployment of Soviet self-propelled artillery units. In addition, by the summer of 1943, the combat experience of both self-propelled artillery crews and unit commanders remained relatively limited. However, despite the need for changes, they were not implemented, and the negative trend persisted after the battle ended and throughout 1943.

On January 6, 1944, the commander of the Tank and Mechanized Forces of the Belorussian Front issued an order titled "On the combat employment of self-propelled artillery," where he explicitly stated that "until this time, tank and combined arms commanders had been using self-propelled artillery improperly. They failed to employ it with effective fire and tactical coordination with tanks, infantry, and artillery. Instead, they positioned self-propelled artillery in front of tanks and infantry battle order, essentially using them as tanks."[104] By this time, it was understood that the issue was widespread, systemic and required immediate reaction.

On February 13, 1944, the new Combat Regulations for Tank and Mechanized Forces (BT and MV) was accepted for service.[105] While self-propelled artillery was mentioned there among other armored combat vehicles, the document didn't provide any further instructions or recommendations on its employment.

It was not until September 1944 that the self-propelled artillery received its own doctrinal document named "Manual on employment and actions for self-propelled artillery." It was largely built upon the lessons learned in combat during 1943–44, offering greater precision and more detailed information compared to the earlier doctrinal documentation.

The emergence of these high-level doctrinal documents was undoubtedly a significant step forward in the development of the theory of Tank and Mechanized Forces. However, their implementation was accompanied by two parallel processes: the massive introduction of the new combat materiel, and evolution of the tactics on both sides. As a result, the higher-level documents were supported with additional orders, instructions, and directives, aimed at clarifying the elements of combat use for self-propelled artillery of different types and models, in different combat scenarios and environments.

Most of these documents were of a temporary nature; some of them became irrelevant with the introduction of new combat equipment or changes in the nature of combat operations. However, a portion of them remained in force simply because there was no adequate replacement for them. For example, in his order released in January 1944 the commander of the Tank and Mechanized Forces of the Belorussian Front mentioned that it should be used "until the combat regulations for self-propelled artillery of the Red Army is issued."[106] This document, however, did not materialize during the Great Patriotic War.

While some features remained a subject of debate until the end of the Great Patriotic War, a consensus on the specifics, combat employment, and tactics of self-propelled artillery was eventually reached in the second half of 1944. The first major doctrinal improvement was the introduction of precise definition for self-propelled artillery. The "Manual on employment and actions for self-propelled artillery" defined it as "any gun of any caliber mounted on wheeled or tracked chassis, serving as its carriage."[107]

The same document provided a classification of self-propelled artillery per weight class—light, medium, and heavy—and combat role. According to the manual, there were four main classes:

a) Direct fire support self-propelled artillery;
b) Anti-tank self-propelled artillery;
c) Anti-aircraft self-propelled artillery;
d) Heavy self-propelled artillery for engaging fortifications, such as pillboxes or concrete bunkers.[108]

The main features of self-propelled artillery remained largely unchanged, encompassing the same combination of firepower, armor protection, and mobility.[109] Some documents, however, added to this combination a larger operational radius and greater operational mobility when compared to towed artillery,[110] and "permanent readiness for opening fire."[111]

The list of combat roles envisioned for self-propelled artillery, in contrast, evolved during the period between 1943 and 1945. For instance, in July 1943, the commander of the Bryansk Front ordered that SPGs should be used primarily for supporting tank units with direct fire. He emphasized that self-propelled artillery should follow advancing tank formations and engage targets from open and half-concealed positions. Using self-propelled artillery to support infantry units was allowed only in exceptional cases. The place of self-propelled artillery was at a distance of 800–1000m behind advancing infantry formations.[112]

By January 1944, the doctrine returned to a broader utilization of self-propelled artillery. The manuals and directives now ordered its use in support of "tank, infantry, and cavalry units." Depending on the tactical situation, self-propelled artillery units could be attached to rifle, tank, or mechanized formations and employed to support attacks, exploit breakthroughs, or serve as a mobile fire support reserve.[113]

Consequently, the list of tasks changed accordingly. Regardless of the weight or class of their self-propelled guns, self-propelled artillery units were responsible for engaging enemy tanks and means of fire support that impeded the advance of allied forces. In defense, it was ordered to use SPGs as "armored fire positions [dug in], in ambushes or as roving guns."[114] Other possible ways of employment included supporting counterattacks by allied forces and serving as an anti-tank reserve. During a retreat, it was permissible to attach SPGs to rearguard units.

The combat formation for self-propelled artillery units was quite straightforward at this point: a one-line formation for a battery, with 40–50m wide intervals between light or medium SPGs and 50–75m between heavy SPGs. For a regiment, it was recommended to use either a single-line or a two-line (two-echelon) formation. In this case, a regiment deployed batteries with intervals of 100–200m between them. The place of a self-propelled artillery regiment was at a distance of 300–600m, depending on terrain and caliber of SPGs, behind the advancing infantry or dismounted cavalry.[115]

By November–December 1944 the doctrine had changed again. Firstly, in offensive operations the place of a self-propelled regiment now was inside advancing infantry or cavalry formations, but behind tank formations. Some manuals specified a maximum interval of 50–75m between infantry and SPGs.[116]

The list of tasks for self-propelled artillery was expanded to include new responsibilities. One of these tasks was mopping up enemy anti-tank guns, anti-tank rifles, and infantry that had survived the initial bombardment. Another task was to provide support for tank, infantry, or cavalry units, during the exploitation of a breakthrough or the widening of the breakthrough area, with artillery barrages. Additionally, self-propelled artillery was assigned the duty of conducting "focused fire strikes" against enemy headquarters and troop concentrations in offensive operations.

In encounter battles, long-range fire missions were assigned as a task to self-propelled artillery. The primary objective of these fire missions was to provide covering fire for vanguard troops, while the secondary objective was to target and disrupt enemy troops on the march, including the destruction of field headquarters and the disorganization of the enemy's command and control system.

In defense, self-propelled artillery was responsible for conducting long-range fire missions from forward positions in order to cover forward detachments defending the obstacle belt—minefields, wire obstacles and vanguard positions.

It should be emphasized that while these tasks were organic to artillery in general, Soviet self-propelled guns were not particularly well-suited for indirect fire or long-range fire missions, primarily due to their tactical and technical characteristics.

Finally, in encounter battles, self-propelled artillery could be integrated in flanking or encircling forces, which were intended to suddenly appear on the enemy's flanks or rear, thereby cutting off and isolating their formations from each other.

Since 1943, the Red Army had faced issues related to mechanical reliability and the effectiveness of repair and recovery units for self-propelled artillery, which likely led to further restrictions. In late 1944, documents were issued that prohibited the self-propelled artillery from being redeployed over long distances under its own power.[117]

Because tactics and equipment frequently evolved, the suggested engagement distance for SPGs underwent continuous adjustments, leading to numerous conflicting recommendations and orders. For instance, the "Temporary Regulations" of 1943 suggested engaging the enemy at distances of "1000 meters and less" without specifying the SPG class and type.[118] An order issued in January 1944 established a direct fire distance of up to 1000m for light SPGs and up to 1500m for medium and heavy SPGs.[119] A directive for heavy SPGs of June 1944 recommended attacking tanks at a distance of 2000m and area targets from up to 3000m.[120] Finally, the "Memorandum for a self-propelled artillery crew" issued in December 1944 provided new information: 600m for the SU-76, 900m for the SU-85, and 1200–1500m or less for the ISU-122 and ISU-152.[121]

Equally debatable during the course of the Great Patriotic War was the question of using self-propelled artillery in different fire modes including indirect fire. In 1942, it was envisioned to use self-propelled guns not only for direct fire support from static positions and short stops but also for firing on the move and indirectly.

Consequently, the requirement for indirect fire was included in the most technical and tactical specifications of self-propelled guns in 1942. While trials confirmed the pre-war experiments and the possibility for using SPGs for indirect fire, the battlefield experience produced mixed results.

The early impressions were mostly positive. For example, the inspection of self-propelled artillery regiments equipped with SU-122, SU-152, and SU-85 SPGs in the autumn of 1943 stated that these units frequently utilized indirect fire capabilities in "all forms of combat" and were capable of conducting such missions not only during the day but also at night.[122]

On the other hand, most of the battlefield reports that emerged in 1944–45 expressed the opposite opinion: self-propelled artillery was rarely used for indirect fire. As stated in one of the reports dated March 1944, "the existing Hertz panoramic sight could easily be replaced with a tank telescopic sight of a TMFD type."[123]

The majority of post-war reports agree that while theoretically it was possible for self-propelled artillery to fire on the move and indirectly, both firing modes were deemed ineffective and used quite rarely.

One of the reports provides a comprehensive explanation: "Indirect fire was used in exceptional cases because the army was saturated with field artillery, and the officer corps of self-propelled artillery lacked sufficient theoretical training. If firing from a concealed firing position was used, it was done against heavily fortified enemy defense lines," i.e. against static area targets.[124] The post-war report of the 367th Heavy Self-Propelled Artillery

Regiment offers more insight into the disadvantages associated with using self-propelled artillery for long-range fire. It noted that firing at distances of 2–5km was not considered advantageous, as observation from the self-propelled gun (SPG) was limited and often obstructed by smoke and dust generated during firing. These factors hindered both fire control and target designation. A distance of 1000–2000m was considered as optimal for engaging enemy tanks and self-propelled guns, whereas attacking soft area targets, such as uncovered infantry, was deemed effective at distances of up to 4000m. At the same time, fire on the move was regarded as impossible. The report stated that attempts to fire on the move from heavy ISU-122 and ISU-152 SPGs led to gearbox breakdowns.[125]

Nevertheless, there is evidence suggesting that although indirect fire was infrequently employed throughout the army, certain units utilized this firing mode more often. For example, in a post-war report authored by the Chief of Artillery Staff and the Commander of Artillery for the 4th Guards Tank Army, it is mentioned that self-propelled artillery frequently utilized this firing mode.

Firstly, it was employed for long-range fire missions at distances of up to 6km at the outset of encounter battles. Secondly, the report mentions that SPGs often carried out indirect fire missions from partially or fully concealed positions, aiming with a panoramic sight. This approach proved highly effective in suppressing individual targets or centers of resistance. However, the report emphasizes a key condition for success: self-propelled artillery needed to be positioned and utilized as ground (towed) artillery. Finally, self-propelled guns were used to fire at distances of up to 6km during offensives to "demoralize enemy rear areas and disrupt lines of communication." The report did not provide further details on the efficiency of long-range fire. Nevertheless, considering the tactical and technical characteristics of Soviet self-propelled artillery, it is safe to suggest that indirect fire was effective for harassing fire or suppression of static or large area targets.[126]

The same report mentions two methods for employing self-propelled artillery at night. The first method involved target designation using flares and forward observers. The second method involved preparing bundles of straw along the anticipated enemy counterattack route. In case of an enemy advance, the designated self-propelled gun crew members were instructed to ignite the straw, illuminating the battlefield and aiding the self-propelled artillery in target acquisition.[127]

The centralized and decentralized use of self-propelled artillery was another area that changed over time. Between 1943 and early 1944, the doctrine not only allowed but also recommended the decentralized use of self-propelled artillery. This concept was first introduced in the "Temporary Regulations" issued in January 1943 and persisted at least until early 1944. For instance, a document from January 1944 permitted the use of even single SPGs in coordination with infantry assault groups during attacks against enemy fortifications.[128]

However, concerns over decentralization were voiced as soon as in late 1943. After inspecting self-propelled artillery regiments of the Western Front, Lieutenant Colonel-Engineer Kostsov from the Self-Propelled Artillery Directorate wrote in his report: "It is not recommended to break up self-propelled artillery regiments; these regiments should be employed as a whole in all forms of combat.[129] The trend continued in 1944, and by November–December the use of single self-propelled guns was strictly prohibited, while breaking up regiments into separate batteries was allowed only as an exception.[130]

In 1945, the practice of decentralization was almost unanimously considered wrong and was officially restricted. However, instances of decentralized self-propelled artillery use were still common, even in the later stages of the Great Patriotic War. For example, the Chief of Staff and Commander of Artillery of the 5th Guards Tank Army highlighted in their report in March 1945 that "there were cases of breaking up self-propelled artillery units into separate batteries and even into single combat vehicles. These vehicles were used on a broad front, without any possibility for maneuvering or establishing deep anti-tank defense with a comprehensive fire system."[131] Although some commanders at the front continued to break up self-propelled artillery units into smaller groups, typically this practice was strongly criticized by army and front-level commanders.

Post-war reports indicate that breaking up self-propelled artillery units was not recommended and should only have been considered when operating in the enemy's operational depth. Another accepted reason for decentralized self-propelled artillery use were combat actions on specific types of terrain, such as forests, swamps, mountains, or in urban warfare.

There were several reasons for such a significant shift in doctrinal thinking. Firstly, maintaining command and control over decentralized units proved problematic. Equally challenging was providing effective fire control and concentrating fire on a single target or adjusting fire. The method of fire concentration was eventually accepted as the most effective for self-propelled artillery.

Another issue was supplying self-propelled guns with ammunition, fuel, and lubricants on the battlefield. Supplying a unit dispersed over a large territory during battle was, and still is, a formidable task. Finally, the decentralized but coordinated use required the highest level of proficiency from everyone in the self-propelled artillery regiment, as well as stable communication between unit command and every crew. Achieving this level of training was, in most cases, unattainable for the Red Army in the 1940s and is rarely found in modern militaries.[132]

Another significant development introduced in 1944 was the definition of specific task and combat roles for different classes of self-propelled artillery. For example, SU-85 guns earned their reputation as effective anti-tank weapons. As a result, in mid-1944, self-propelled artillery regiments armed with SU-85 SPGs were included in the anti-tank artillery brigades (IPTABr). These regiments were considered an "organic and inseparable part of anti-tank artillery brigades," specifically intended for destroying "enemy tanks and self-propelled guns." Directives and manuals issued during this period explicitly prohibited using SAPs for other tasks or breaking them into parts.[133]

Another manual issued in 1945 mentioned that SU-76 should be used mainly for supporting infantry formations. Accordingly, regiments armed with SU-76 SPGs were considered as "infantry support artillery that is less vulnerable to enemy fire and more mobile compared to horse-towed regimental artillery."[134] It is worth noting that by 1945 the 76mm ZiS-3 gun, main armament of the SU-76 SPG, was not seen as an effective anti-tank weapon, thus the role of these lightly protected combat vehicles was largely limited to direct infantry fire support.[135]

Finally, during 1944, the combat roles and tactics for heavy self-propelled artillery were gradually defined and formalized. On May 29, 1944, a document emerged that provided clarification regarding the combat employment of tank regiments equipped with IS-122 (IS-2) heavy tanks and self-propelled artillery regiments armed with ISU-152 heavy SPGs. The instructions highlighted that the IS-2 and ISU-122 were considered "the best means of armed combat" against heavily armored German tanks and self-propelled guns, including the Panzerkampfwagen VI Tiger, Panzerkampfwagen V Panther, and Sd.Kfz. 184 Ferdinand, as well as enemy artillery and heavy fortifications. Generally, heavy tank and self-propelled artillery regiments were viewed as force multipliers at the corps level. It was envisioned to attach 1–2 regiments to each tank corps, mechanized corps, or, in certain cases, rifle corps.[136]

Another document, dated June 12, 1944, provided more details about the tasks of heavy Guards tank and self-propelled artillery regiments. These tasks included "engaging in combat with heavy tanks and self-propelled guns and countering anti-tank artillery and fortifications."[137]

It was recommended to engage the enemy with direct fire from a distance of up to 2000m. However, the instructions emphasized that the most effective fire could be provided from distances of 1500m and less. The document highlighted that "fire from IS and ISU at a distance of 1500m or closer can destroy any enemy heavy tank. It is especially advantageous when you can approach the enemy tanks from the side."[138] However, the directive issued in July 1944 allowed to engage area targets, such as artillery positions or troop concentrations, from a distance up to 3km.[139]

One of the essential elements for successful employment of self-propelled artillery or heavy tank regiments was screening heavy tanks or SPGs with infantry, medium tanks and lighter SPGs. They were meant to attack or move in the first line, while heavier vehicles, with their slower reloading cycle, were supposed to identify targets and engage them from longer distances. This typically meant they operated in the second line (echelon), from the rear of formations armed with light SPGs or medium tanks.

The same scheme applied to movement and maneuver on the battlefield, where heavy tanks and self-propelled guns were expected to follow infantry or medium tanks at a distance of 300–500m, moving "from one position to another, combining fire and maneuver." It was recommended to fire while stationary or from short stops, utilizing open, half-concealed, or fully concealed firing positions.

Following close to infantry or medium tanks battle formations was deemed as an essential element for maintaining good situational awareness, while lagging behind was considered as a main reason for "not engaging enemy in a timely manner." It was also considered essential to cover flanks for heavy tank or self-propelled artillery regiments with "medium tanks or infantry supported by artillery." Additionally, the instructions emphasized that "every heavy tank or heavy self-propelled gun must have submachine gunners for cover."[140]

The typical combat formation recommended for heavy self-propelled artillery regiments was a two-line (echelon) formation, with 2–3 batteries in the first line and 1–2 in the second. Batteries on the left and right flanks were positioned in the rear echelon formation to cover the flanks. The intervals between companies in the same line were set at 100–200m, and the intervals between individual vehicles were 50–75m. The distance between the first and second lines (echelons) was set at 200–400m.[141]

In offensive operations, the primary objective for heavy tank and heavy self-propelled artillery units was "destroying enemy tanks and self-propelled guns, and repelling enemy counterattacks." In certain cases, like storming heavily fortified enemy positions, it was allowed to use heavy SPGs in the first line for destroying concrete fortifications and bunkers. In these situations, tanks or SPGs were attached to assault groups of the rifle regiments.[142]

In defense, heavy tanks and SPGs were intended to function as powerful means of anti-tank defense along the main axis of an enemy attack or as a mobile anti-tank reserve. A heavy tank or heavy self-propelled artillery regiment was supposed to take a defensive position in a chessboard pattern and was capable of covering a sector 2.5km wide and 2.0km deep. The directive recommended using covers or concealed positions whenever possible and preparing at least 2–3 reserve positions for each tank or SPG. Additionally, the instructions ordered positioning tanks or self-propelled guns to "establish a mutual fire support system among themselves and anti-tank field artillery."

A typical defensive system involving a heavy tank or self-propelled artillery regiment was described as follows: "The first line consists of infantry, artillery, and medium tanks. Behind them, at a distance of 500–600 meters, 2 or 3 companies of IS tanks or ISU self-propelled guns should be deployed. The remaining tanks [SPGs] of the regiment should be deployed at a distance of 300–1000 meters and function as a mobile reserve."

In defense or ambush, enemy heavy tanks and heavy self-propelled guns were designated as primary targets for Soviet self-propelled artillery.[143] It was strictly prohibited to deploy IS-2 heavy tanks and ISU-152 self-propelled guns in front of the medium tanks or infantry formations and to use them as "static guns."[144]

The majority of manuals and instructions issued in 1944–45 agreed that probably the most effective tactic for self-propelled artillery was ambush. The recommendation was to deploy SPGs along the probable directions of enemy attacks, with the aim of luring enemy tanks into close range and destroying them with precise, sudden fire.[145]

Observations can be made from manuals for heavy tank and self-propelled artillery units. Firstly, there is no significant difference in tactical employment between heavy tanks and heavy SPGs. Secondly, both heavy tank regiments and heavy SPG regiments were clearly regarded as high-value assets, requiring numerous preparatory actions and supporting elements to be in place. Lastly, the primary objectives for heavy tanks and self-propelled artillery were enemy armored vehicles, anti-tank artillery, and fortifications, contrary to some works suggesting their main adversaries was German infantry.[146]

Contrary to common perceptions regarding the Soviet heavy tanks and self-propelled guns as means of breakthrough spearheading offensives, these were often used as heavy artillery in direct fire support role. In certain cases, heavy tanks and SPGs were used together to provide artillery fire support for breakthroughs. The combat report of the 11th Tank Corps mentions an episode when an IS-2 tank regiment was combined with a heavy ISU-152 regiment to support infantry during the breakthrough. The report emphasizes that "because towed artillery got stuck in the mud, and supply with wheeled transport was impossible, heavy tanks and self-propelled guns were the only artillery available to support the offensive actions."[147]

Another common tactic that emerged during 1944–45 was using self-propelled artillery with assault (shock) groups. As the war progressed and the Red Army shifted toward offensive operations, it faced a growing need to storm fortified areas, positions, and cities. Mobile and well-armored self-propelled guns equipped with powerful cannons proved to be extremely effective in urban combat, in forested or mountainous terrain, and as a means of destroying enemy fortifications.

Actions with assault groups were probably the only combat scenario where the doctrine allowed breaking up regiments and batteries and using single vehicles. The reason for this was obvious: the congested combat environment rarely permitted the deployment of large units in full strength.

The first mentions of this tactic could be traced to early 1944. The documents of this period mention that individual SPGs, preferably armed with heavy-caliber guns, should be employed in attacks against fortified areas. Their task was to block or destroy heavy enemy fortifications, such as bunkers or pillboxes in cooperation with assault (shock) groups.[148]

The tactics of assault groups evolved over time and became widespread across the army during 1944–45. In September 1944, the tactic was described and formalized in the "Manual on employment and actions for self-propelled artillery." The document suggested using individual vehicles or SPG batteries for destroying or suppressing enemy fortifications that survived after the preparatory artillery barrage or airstrikes.[149]

The recommended course of action involved engaging enemy positions with direct fire from open or half-concealed positions. The key conditions for a successful assault included thorough reconnaissance and planning, establishing close cooperation with infantry, sappers, and other supporting elements, setting up communication and intelligence-sharing channels with supporting units (tank, infantry, and artillery), and, lastly, conducting rehearsals and drills with the assault troops before launching the attack.[150]

Urban combat was a challenging environment for command and coordination that limited maneuverability and restricted observation. For these reasons, self-propelled guns in urban combat were typically employed in small groups, batteries, pairs, or even individually.

Groups typically consisted of 2–4 SPGs due to the severe constraints of the urban environment, allowing only 2–3 forward-deployed vehicles to fire simultaneously. The typical engagement distance ranged from 150 to 300m. Destruction and suppression fire were carried out in short salvos, involving 3–4 SPGs, at a "minimum possible distance, as enemy anti-tank infantry fire allows."[151]

Self-propelled artillery employed a technique known as the leapfrog method for advancing and moving. One group or battery would provide cover and suppressive fire while the other element moved forward to the next position.

Similar to combat in fortified areas, thorough reconnaissance was a key prerequisite for success. Another defining factor was the inclusion of infantry alongside self-propelled guns. Assault groups and combat vehicles mutually supported each other: self-propelled artillery provided transportation, radio communications, and formidable firepower to the accompanying infantry. Simultaneously, assault troops shielded SPGs from German anti-tank infantry and provided SPG crews with target designation and intelligence, thereby enhancing situational awareness.

The report of the 399th Heavy Self-Propelled Artillery Regiment describes an episode from the Battle of Berlin in which a group of assault infantry advanced ahead of ISU-122 SPGs. When the infantry spotted two concealed German anti-tank guns, they started firing tracers in their direction, designating targets for the SPG crews. One of the ISU-122s began suppressing the anti-tank guns with heavy machine gun fire and then swiftly advanced, crushing the guns with its tracks.[152]

The importance of suppressive machine gun fire was one of the key lessons learned from urban combat. Many reports emphasized the value of 12.7mm heavy DShK machine guns installed on self-propelled guns, such as the ISU-152 or ISU-122. Some reports even recommended firing on the move to the windows and possible hiding places on the streets in front of advancing vehicles.[153] However, in this case, the machine gunners were exposed to enemy fire from rooftops and upper stories of buildings. Soviet units that participated in urban combat frequently reported this issue. For instance, a report from the 367th Heavy Self-Propelled Artillery Regiment noted that while DShK heavy machine-gun mounts played a crucial role in fighting against enemy aircraft and infantry, the SPG commanders had to expose themselves up to their waists, making them vulnerable to enemy fire. It's important to mention that the 367th Regiment was equipped with heavy ISU-122 and ISU-152 self-propelled guns. These SPGs had armored casemates, offering protection against top attacks. The vulnerability to such attacks was even more pronounced in open-topped vehicles like the SU-76 and T48.[154] Modern experience of urban combat largely confirms the need for armored vehicles to be equipped with a heavy machine gun in a protected or remotely controlled mount.

Both wartime and post-war reports generally agreed that self-propelled artillery, regardless of caliber, achieved excellent results when integrated with assault groups, yielding a "great effect." However, these positive outcomes were contingent upon the correct employment of self-propelled artillery units and adherence to doctrine, which did not always occur.

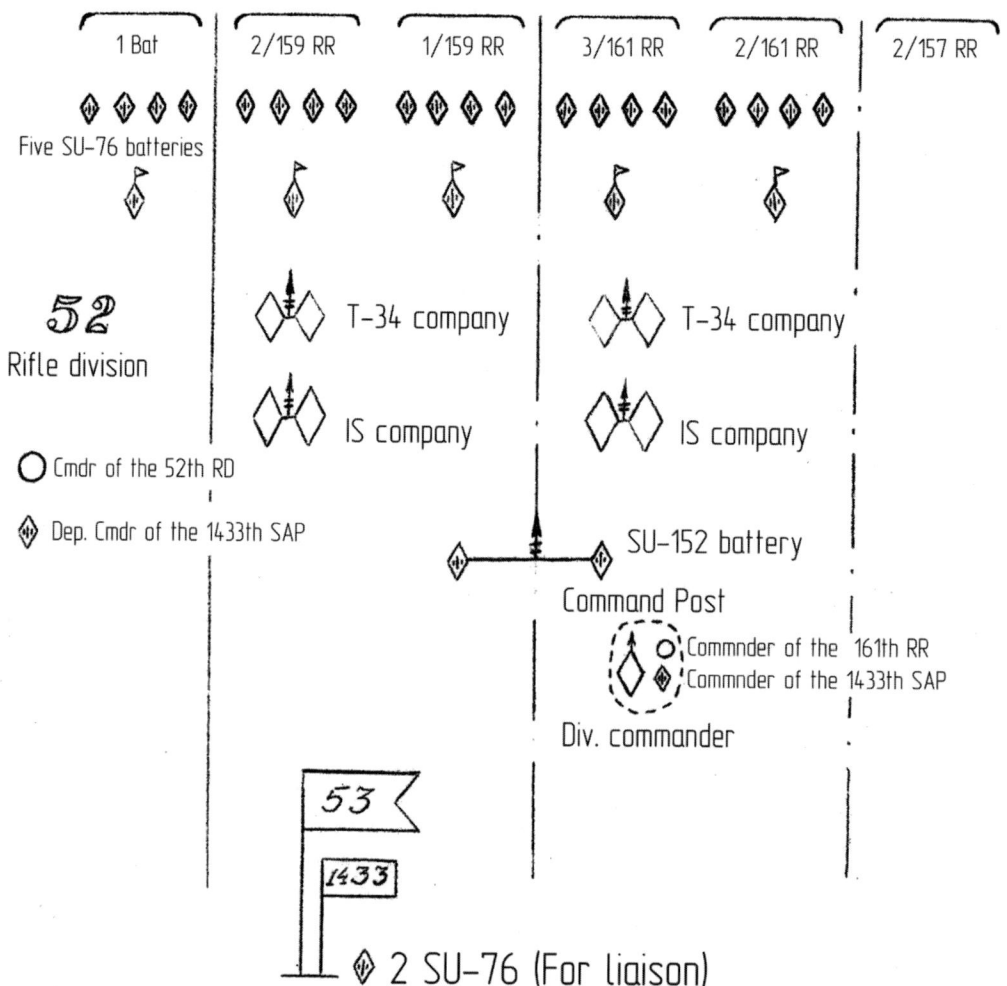

Battle Formation Scheme for Breaking through the Fortified Zone near the town of Laura. August 10, 1944. Note that the formation is led by rifle regiments and battalions with support from SU-76 light self-propelled guns. The heavy IS tanks and ISU-152 self-propelled guns (SPGs), protected by medium tanks, are positioned in the 2nd and 3rd lines.[155]

Chapter 5

In the Flames of Battle

The first combat employment of a new weapon or an entirely new branch or arm of the military represents the moment when the best technological advancements, achievements in doctrinal thinking, training, and overall military organization converge. In theory, all these elements are poised to work cohesively as a whole to achieve the desired outcome on the battlefield.

However, the military system is sensitive to any flaws. Deficiencies in one of the key elements—materiel, personnel, training, or organization—inevitably affect the performance of the arm of service and, broadly, the entire army. This was indeed the case with Soviet self-propelled artillery. When employed correctly in accordance with doctrine and manuals, it made a significant impact on the battlefield. "Self-propelled artillery regiments are modern and the most effective means of support for a tank unit in combat. With their ability to deliver powerful fire support while moving alongside tanks, they provide tank units with long-range fire and increase their survivability." This high, positive appraisal was given to self-propelled artillery in August 1943, after the Battle of Kursk.[156]

However, the cases when self-propelled artillery was employed completely by the book were rare. Typically, the employment in combat suffered from shortcomings in different aspects of the military system or tactical errors. As self-propelled artillery evolved and gained combat experience, some shortcomings were fully or partially resolved, while others persisted throughout the Great Patriotic War and continued to exist even after its conclusion. Perhaps the most significant and persistent problem was the widespread use of self-propelled artillery for roles not suited for it doctrinally or by design, such as employing them as tanks or assigning them unusual and inappropriate tasks.

Although the doctrine explicitly restricted the use of self-propelled guns as tanks and emphasized the need to distinguish between these two distinct classes of armoured fighting vehicles, an analysis of their initial deployment in 1943 revealed that self-propelled guns were, in practice, employed in a wide range of combat and auxiliary tasks.[157] In July 1943, the Commander of the Bryansk Front, in his combat order, mentioned instances in which self-propelled artillery was used as infantry support tanks, as well as for attacking enemy positions independently, without additional support.[158]

Another document describing the actions of self-propelled artillery regiments on the Central Front in July 1943 mentions that "there were instances when self-propelled guns were used alongside tanks or, in some cases, even in front of tank formations. In such a battlefield deployment, SPGs essentially did not differ in their application from tanks."[159]

The Commander of Tank and Mechanized Forces of the Belarusian Front, in his order regarding the combat deployment of self-propelled artillery issued on January 6, 1944, explicitly states, "Tank and combined arms commanders often misuse self-propelled artillery, employing it without proper fire or tactical cooperation with tanks, infantry, and artillery. They place it in front of tank and infantry combat lines, essentially using self-propelled guns (SPGs) as if they were tanks." He goes on, providing an example. On December 5th, 1943, the commander of the 397th Rifle Division ordered to support an infantry attack with 152mm self-propelled guns. The SPGs followed the order and advanced alongside the first line of infantry. The outcome of the 30-minute combat was discouraging: four SU-152s were lost without achieving any significant results.[160]

Identical practice was observed during the inspection of self-propelled artillery units on the Western Front in the autumn of 1943. The inspection revealed that self-propelled artillery was often used for tasks not considered in the doctrine or even specified in the tactical and technical requirements. The report stated, "Currently, self-propelled artillery is employed in various forms of combat, including offensive and defensive operations, encounter battles, escorting and providing security for tank, motorized, and infantry columns on the move, providing support for avant-garde detachments, securing river crossings, as well as delivering indirect fire support in all types of combat actions, and for equipment recovery."[161]

In his report, Lieutenant Colonel-Engineer Kostsov specifically identified eight major scenarios, where self-propelled guns (SPGs) were employed incorrectly:

1. In offensive actions as tanks;
2. For reconnaissance without any support;
3. For infantry support without any assistance;
4. For performing tank tasks when tanks were available;
5. In a first attack line (echelon) for engaging enemy with tracks, and not in a second echelon for fire support;
6. As APCs for transporting troops;
7. As tractors for damaged materiel recovery and hauling;
8. For attacking enemy fortified positions without any support.

Kostsov emphasized that improper use of self-propelled artillery always led to significant losses and resulted in combat mission failure. He recommended refraining from employing SPGs for tasks they were not designed for.

It's worth noting that Kostsov inspected seven regiments. Three of them (the 1537th, 1494th, and 1833rd) were armed with heavy SU-152 self-propelled guns; another three (the 1830th, 1819th, and 1435th) were equipped with medium SU-122 SPGs. The last regiment was armed with SU-85 SPGs, the newest self-propelled guns at that time.[162] In other words, the deficiencies in combat use were not determined by class, model, or intended combat role; rather, they were rooted deeper.

Multiple instances of incorrect use of self-propelled artillery in 1943 could be attributed to a lack of experience. However, the situation did not significantly improve in 1944–45. On January 6, 1944, the Commander of Tank and Mechanized Forces of the Belorussian

Front reported that, "Until this point, tank and combined arms commanders had been misusing self-propelled artillery, deploying it without fire or tactical coordination with tanks, infantry, and artillery. They were placing it in front of tank and infantry formations, essentially using self-propelled artillery (SPGs) as tanks."[163]

The same issue was raised in another report that surfaced ten months later, on November 17, 1944. The Commander of the First Baltic Front noted, "During the military operations of the Front in September and October 1944, there were instances of incorrect employment of self-propelled artillery regiments (SAPs) by the commanders of combined arms and artillery units and formations, to which these regiments were assigned as a means of reinforcement."

Further, he pointed out the major issues related to combat employment of self-propelled artillery units. Firstly, self-propelled artillery regiments were assigned tasks as if they were regular tank units, with little consideration for their combat capabilities and the tactical and technical characteristics of their equipment.

Secondly, self-propelled artillery units in combat were not provided with adequate infantry support. Conversely, infantry units structured their combat formation as if they were operating with tanks, allowing a significant distance (over 200m) between self-propelled guns and infantry.

Thirdly, self-propelled artillery units were often employed in a piecemeal fashion. There were instances when single vehicles were deployed across a wide front, leading to significant challenges in commanding and controlling the actions of the entire unit.

Finally, self-propelled artillery regiments were deployed in battle without prior coordination with infantry and artillery units, and without conducting any reconnaissance of the terrain or enemy. Radio communications were only established once the battle had commenced, frequently resulting in higher-level combined arms commanders losing control over the self-propelled artillery units in the midst of battle.[164]

Many instances of improper use of self-propelled artillery were documented between November 1944 and May 1945, persisting throughout this period. It is worth noting that this problem existed at all levels, from brigade to army, despite the fact that by 1944, the Red Army had accumulated sufficient combat experience with mechanized units and was under less pressure than in 1943 and the first half of 1944.

A report on the actions of towed and self-propelled artillery of the 5th Guards Tank Army in January–February 1945 vividly illustrates instances of incorrect utilization of both artillery categories, despite their intended roles in supporting one another and the operations of this high-level tank formation. Firstly, the report noted the cases when single self-propelled guns were used for battlefield reconnaissance.[165] Secondly, tank commanders assigned tasks to self-propelled artillery regiments as if they were tank regiments, without considering their combat capabilities, technical, and tactical features, and used them as "attacking tanks." At the same time, the organic tasks of self-propelled artillery were frequently assigned to towed artillery.[166]

Ironically, both organic and attached towed artillery units were also employed improperly. The report specifies that tank unit officers often assigned tasks to the organic anti-tank batteries of tank and motor-rifle brigades. However, in reality, "nobody was in command of these units. During rapid offensives, anti-tank batteries were frequently left behind messing around in the rear. In the best-case scenario, they were assigned to secure brigade headquarters."

Furthermore, both corps and army-level artillery attached to the tank and motor-rifle brigades were often used for direct fire support. According to the report from the 5th Guards Tank Army, corps artillery was deployed for direct fire, with instances where entire regiments were deployed for this purpose. Notably, corps artillery was deployed in this manner 'sometimes without any necessity, intelligence about the enemy's fire system, and without any reconnaissance." Finally, some tank commanders tended to use army artillery as "accompanying artillery for direct fire support," neglecting the capability of heavy artillery to create depth in artillery fires or to place concentrated fire or moving artillery barrages on potential enemy attack axes.[167]

Another mention of the same problem can be found in the report of the 31st Tank Corps, dated December 17, 1945, which states: 'self-propelled artillery was frequently used within tank formations, not as a force multiplier for engaging enemy fortifications, but rather as tanks, resulting in unnecessary losses."[168]

Generally, many wartime and post-war reports and analyses, based on battlefield experience and written by veteran commanders of tank and mechanized troops, acknowledge the issue of improper employment of self-propelled artillery at all levels.

The core problem and its consequences were formulated in the report on artillery support of the tank corps presented at the conference of the 5th Guards Tank Army in December 1945. The section on the Battle of Prokhorovka stated: "This battle demonstrated the significance of cooperation between all arms. During this battle, the 1446th Self-Propelled Artillery Regiment suffered heavy losses simply because nobody knew how to employ it effectively. Self-propelled guns were positioned not behind the tank formations but in front of them, because they were ordered to do so."[169]

It's worth noting that assigning unconventional tasks to self-propelled artillery units did not always occur due to incompetence or errors. Sometimes, it was a deliberate response to specific circumstances, like the tactical situation or equipment shortages. The report from the 31st Tank Division serves as an example of this. In February 1945, due to an insufficient number of combat-ready tanks at hand, a self-propelled artillery unit was assigned to act as a tank unit and launch an attack on Wolfskirch (now Wilczkowice) village near Breslau. The village was heavily fortified with German anti-tank infantry teams and artillery. The attack resulted in self-propelled unit suffering heavy losses and subsequent retreat without accomplishing the mission.[170]

Another example can be found in Lieutenant Colonel-Engineer Kostsov's post-inspection report. He notes that during the periods of intense fighting, self-propelled guns often use 3 to 5 full combat loads of ammunition per day. The ammunition capacity of self-propelled guns was considered inadequate. But as there were no dedicated armored ammunition carriers, the Red Army had to improvise. Tankers and self-propelled artillery crew members resorted to using horse-drawn carriages, unarmored trucks, or even loading additional ammunition crates onto or inside the self-propelled guns.[171]

The problem of improper combat employment remained largely unresolved until the end of the Great Patriotic War. The underlying reasons and the ultimate outcome remained consistent with what had been identified in 1943: a lack of sufficient knowledge across the officer corps, and inadequate experience on how to employ self-propelled artillery correctly. While these deficiencies generally did not result in large-scale operational failures, they often caused significant and unnecessary losses in both personnel and equipment.

Another significant problem was the lack of cooperation and interoperability between self-propelled artillery and both organic and allocated supporting elements. The importance of cooperation between branches of service runs as a common theme in many doctrinal documents issued between 1943 and 1945. Observations made in 1943, following the Battle of Kursk and offensive actions at the Bukrin Bridgehead, emphasize two crucial requirements for achieving success: correct employment of self-propelled artillery and established cooperation with other arms of service.[172]

In 1944, this requirement was included in instructions and manuals. For example, the Manual on employment and actions for self-propelled artillery defines it as a matter of "exceptional importance."[173] This perspective dominated throughout the war, and by the end of the Great Patriotic War, Soviet military theory had come to the unified concept of modern combat as the "combined effort of all branches of service," with cooperation among them seen as a cornerstone of success in battle. This cooperation, in turn, was viewed as the "cohesive action of allied forces."[174]

While the theoretical basis steadily evolved, its practical implementation on the battlefield posed challenges. For instance, between September 17 and 18, 1944, the 1051st Self-Propelled Artillery Regiment was engaged in combat with the 43rd Anti-Tank Artillery Brigade near Dobele, Latvian SSR. In only two days, the 1051st SAP irrecoverably lost up to 90% of its self-propelled guns. According to the 1051st SAP's War Diary, out of the 16 SPGs (three batteries of five vehicles each, plus one SPG in the staff detachment) that participated in the actions on September 17, twelve were completely lost due to burning out, while the remaining four were damaged and later recovered from the battlefield.[175] The investigation initiated in the aftermath revealed the reasons for these disastrous losses. Firstly, the regiment was engaged in combat on an 18km-wide sector without established communication and cooperation between the self-propelled artillery batteries and the elements of the 9th and 71st Rifle Divisions. Secondly, the rifle divisions failed to provide infantry cover for the self-propelled artillery and the elements of the 43rd Anti-Tank Artillery Brigade. As a result, German forces utilized rugged, forested terrain to bypass and outflank self-propelled artillery batteries, subsequently encircling and destroying individual vehicles.[176]

The scale of the problem is also evident in the aforementioned example describing the actions of the 5th Guards Tank Army in January–February 1945. First, it highlights the challenge of effectively using both self-propelled and towed artillery within a tank army, which, by its very nature, functions as an independent operational-level formation where all elements must operate cohesively. Secondly, it brings attention to the issue of insufficient training, especially prevalent among tactical brigade-level commanders who seemed to lack essential knowledge. In other words, tank commanders often didn't grasp the nature and functions of corps- and army-level artillery, which typically involved providing indirect fire and deployment in the rear. As a result, they were unable to employ heavy artillery correctly and coordinate their actions with artillery units.

Equally significant is that these deficiencies were documented in January–February of 1945. By that time, the 5th Guards Tank Army was already an experienced and battle-hardened formation, established in January–February 1943 and having participated in numerous battles since. Conversely, when cooperation between all elements was established and self-propelled artillery units were employed according to doctrine, the results were positive.

Illustrative examples can be found in the documents that describe the actions of the 29th Tank Corps of the 5th Guards Tank Army during the offensive operations in Northern Poland and Eastern Prussia. As shown in Table 11, just prior to the operation, the 29th Tank Corps had a well-balanced composition that included all the necessary elements for an assault, along with supporting elements.

Table 11. 29th Tank Corps order of battle on January 17, 1945.[177]

Unit	Materiel / Personnel	Strength, on January 17, 1945
31st Tank Brigade	T-34 tanks	39
	SU-76 SPGs	10
	ISU-152 SPGs **	18
	Submachine gunners	255
25th Tank Brigade	T-34 tanks	32
	ISU-152 SPGs **	15
	Submachine gunners	347
32nd Tank Brigade	T-34 tanks	39
	SU-76 SPGs	8
	SU-85 SPGs	23
	Submachine gunners	255
53rd Motor-Rifle Brigade	SU-57 SPGs	8
	APCs	20
	120mm mortars	3
	76mm guns	13
	82mm mortars	30
	Riflemen	695
	Submachine gunners	175
1223rd Light Self-Propelled Artillery Regiment	SU-76 SPG	18
271st Mortar Regiment	120mm mortars	22

Unit	Materiel / Personnel	Strength, on January 17, 1945
165th Light Artillery Regiment	76mm guns	20
409th Separate Guards Mortar Division	M-13 MLRS	8
75th Separate Motorcycle Battalion *	M4A2 tanks	10
	Motorcycles	41
	APCs	11
193rd Separate Sapper Battalion	Sappers	240
366th Anti-Aircraft Regiment	37mm guns	22
	12.7mm DShK HMG	13

* Strength provided on 15 Jan 1945
** The 332nd Heavy Self-Propelled Artillery Regiment was included in the 31st Tank Brigade, while the 365th Heavy Self-Propelled Artillery Regiment was placed within the 25th Tank Brigade.

There are two notable aspects in this order of battle. First, by January 15th, 29th Tank Corps was significantly reinforced with self-propelled artillery. In comparison, in October 1944, the same unit only included the 1446th Self-Propelled Artillery Regiment with 17 SU-85 SPGs and the 1223rd Light Self-Propelled Artillery Regiment with 10 SU-76 SPGs. In terms of quantity, the number of self-propelled artillery vehicles (excluding eight M-13 rocket launchers) increased from 27 to 112, more than quadrupling. With 130 operational tanks, the ratio of tanks to self-propelled artillery was 1:0.86, very close to a 1:1 ratio.

Secondly, self-propelled artillery of the 29th Tank Corps now included all types of SPGs. In addition to the previously existing SU-76 and SU-85 SPGs, the corps received two heavy self-propelled artillery regiments, the 365th and 332nd, equipped with ISU-152 SPGs, as well as a detachment of SU-57 (T48) half-track SPGs.

This represented not only a quantitative but also a qualitative improvement, as the introduction of new types of self-propelled guns (SPGs) dramatically increased the unit's firepower and enhanced the overall combat capabilities of the formation.

More importantly, following the reorganization, self-propelled artillery was integrated into the tank and motor-rifle brigades as organic means of fire support. Two brigades, the 31st and 25th, received heavy ISU-152 SPGs; the 32th Tank Brigade incorporated SU-76 and SU-85 SPGs, while the 53rd Motor-Rifle Brigade's towed artillery was bolstered with SU-57 SPGs.

Table 12. 29th Tank Corps combat strength on January 17, 1945 and before the operation.[178]

	On 17 January 1945	Total strength, before the operation*
T-34 tanks	110	120
ISU-152 SPGs	33	42
SU-85 SPGs	23	25
SU-76 SPGs	36	37
M4A2 tanks	10	10

	On 17 January 1945	Total strength, before the operation*
SU-57 SPGs **	8	3
76mm guns	33	41
57mm guns **	0	7
37mm guns	22	28
120mm mortars	25	25
82mm mortars	30	38
Armored personnel carriers	31	31
Motorcycles	41	41
Riflemen, active	695	659
Submachine gunners	1032	1075

* The materiel that had lagged behind during the march was arriving throughout the course of January 17th, resulting in the following updated totals.
** It appears there might be a typo in the document. The SU-57 SPGs were listed as 57mm guns.

The offensive began on January 17, 1945 with massive preparatory artillery fires and airstrikes that targeted enemy defensive lines and the close rear. After that, infantry units supported by tanks and self-propelled artillery went into action.

Once the rifle units, with designated support, breached the enemy defenses across the entire tactical depth and an 8–10km-wide front, the command deployed the tank and mechanized corps to exploit the breakthrough. During this phase, the artillery was tasked to secure the flanks of the breakthrough area, providing interdiction fire and destroying or suppressing targets that hindered maneuvers of tank formations. The organic artillery of tank and mechanized corps, as well as allocated artillery units, was partially relocated to the flanks of the breakthrough area, and partially moved into the breach with tank formations.

During the maneuver phase, towed and self-propelled artillery units followed the tank units. When tanks encountered enemy resistance, the supporting artillery repositioned themselves in front of the tank combat lines and engaged targets with direct fire, thereby assisting in repelling enemy counterattacks and providing cover for the tactical maneuvers of the tank units. This kind of tank-artillery cooperation enabled tanks to outflank and encircle the enemy, while self-propelled artillery engaged anti-tank and anti-personnel targets. In turn, tanks engaged enemy infantry with their firepower and tracks, while also countering incoming reserves.[179]

On January 17, 1945, the 29th Tank Corps received the order to exploit the breakthrough and advance with the initial objectives of reaching and breaching the enemy's Ciechanów-Przasnysz defensive line, bypassing the fortified area around Mława, and capturing Działdowo by January 18, 1945. The formation advanced in two groups using two routes. The groups had approximately equal strength, with the 31st and 32nd Tank Brigades leading the advance of the first group, and the 25th Tank Brigade and the 53rd Motor-Rifle Brigade spearheading the second group's advance.

On January 21, 1945, the forward elements of the 29th Tank Corps encountered strong enemy resistance on the line of Naguszewo—Gutowo—Zwiniarz—Linowiec, near the city of Deutsch-Eylau. This city served as a vital logistical hub, with communication lines extending into Central Germany. Recognizing its strategic significance, the Germans deployed a substantial force to defend it. According to the War Diary of the 29th Tank Corps, the enemy grouping consisted of elements from the 18th Panzergrenadier Division, and the 23rd and 7th Infantry Divisions, supported by tanks from the 16th Tank Division and an armored train. Furthermore, the city was well-fortified and shielded by natural terrain to the east and south.

The command of the 29th Tank Corps made the decision to eliminate the enemy forces in the Zwiniarz area and to secure the towns of Lubawa and Grabowo, located to the east and southeast of Deutsch-Eylau, by 03:00 on January 21st. Assault and an outflanking maneuver resulted in securing the town of Lubawa by 10:00 on January 21st. By the next day, the advancing elements of the 29th Tank Corps reached the Deutsch-Eylau from three directions: 25th Tank Brigade from the south-east, 32th Tank and 53rd Motor-Rifle Brigades from the west, while the 31st Tank Brigade blocked the city from the east and north-east.

This maneuver placed the defending German forces in danger of complete encirclement. The German forces made four counterattacks in an attempt to break through the Soviet lines and retreat to the west but were unsuccessful. On January 22, 1945, following a fierce two-hour battle, Soviet forces successfully secured the city. According to the documents of the 29th Tank Corps, the garrison suffered irrecoverable losses of over 1500 men and officers.[180]

The documented evidence of combat employment of self-propelled artillery of the 29th Tank Corps allows us to draw some conclusions. Firstly, it unequivocally reaffirmed the pre-war concept of artillery capable of accompanying fast-moving mechanized units into battle and providing direct fire support. The key factor in the success of the operation in Eastern Prussia and Northern Poland was the speed of advance. As described in the report of the 31st Tank Brigade of the 29th Tank Corps, "the lightning pace of the advance prevented the enemy from using their tactical and operational reserves. It also hindered the enemy from occupying their prepared defensive lines, which included anti-tank ditches, barricades, dragon's teeth, and barbed wire." In just six days, the 31st Tank Brigade covered a distance of over 300km, maintaining an average pace of 50km per day, with peak speeds reaching up to 80km per day.[181]

The inclusion of self-propelled artillery in the ranks of tank and motor-rifle brigades greatly contributed to the favorable outcome, as advancing tank units could consistently rely on the overwhelming firepower of the accompanying self-propelled guns. In addition, this organization facilitated swift communication and control between the brigade commander and the tactical commanders in the self-propelled artillery units. Such communication was essential for coordinating the actions of fast-moving troops. Remarkably, the organizational structure of the 31st, 32nd, 25th Tank Brigades, and the 53rd Motor-Rifle Brigade closely resembled the structure of independent mechanized formations introduced in the late 1920s by prominent Soviet military thinkers, such as M. Tukhachevsky, V. Triandafillov, and K. Kalinovsky.

For example, the 31st Tank Brigade included three tank battalions as a "strike core" (a term coined by Triandafillov in 1929), mortars, towed and self-propelled artillery, and mechanized infantry to provide cover for tanks and SPGs and secure the ground. The combat troops were supported by signal, reconnaissance, anti-aircraft, and sapper companies.[182]

There were, however, negative aspects in this organization. Firstly, the distribution of artillery among the brigades did not allow for its centralized employment by the corps commander. Consequently, this prevented him from concentrating firepower on the main axis of advance. Another negative aspect was the difference in speed between heavy and medium vehicles.

For instance, the ISU-152 heavy self-propelled guns were integrated into the organizational structure of the 31st and 25th Tank Brigades, both of which were equipped with T-34 medium tanks. During combat operations, the difference in design and mobility of these platforms presented some challenges. Firstly, the average speed of the ISU-152 was significantly lower compared to that of the T-34. Well-trained drivers were able to maintain the pace of advancing medium tanks, but at the cost of rapid wear and tear on the running gear and ball bearings, which resulted in mechanical breakdowns. Secondly, while T-34 tanks could fire on the move, the design of the ISU-152 self-propelled guns did not allow for this firing mode. As a result, heavy self-propelled guns often lagged behind the advancing tanks by a distance of 3–5km, and even greater distances during the pursuit of retreating enemy forces.[183] To prevent this problem in the future, the artillery command of the 29th Tank Corps proposed equipping mechanized units with uniform platforms, where heavy tanks would be supported by heavy self-propelled guns, and medium tanks with self-propelled artillery on a medium platform.

It's worth noting that in this case, the command of the 29th Tank Corps, based on real combat experience, essentially reverted to the principle of uniformity formulated by Vladimir Triandafillov in his report titled "On the System of Tank-Tractor-Automotive-Armored Armaments of the Red Army" dated June 5, 1929.[184] He wrote, "The main principle, which must lie at the basis of the organization of mechanized formations, is their uniformity in terms of operational and cross-country mobility."[185]

While in most cases self-propelled artillery was at hand supporting tanks, the difference in mobility proved to be more problematic for towed artillery supporting mechanized units. Documents from the 31st Tank Brigade of the 29th Tank Corps mention that the 165th Light Artillery Regiment, armed with towed 76mm guns, "most of the time lagged behind the tanks and failed to provide effective fire support during engagements with enemy tanks and artillery, resulting in unnecessary tank losses."[186]

Another downside that affected tactical and operational mobility was the lack of logistical and auxiliary vehicles. The problem was typical for the Red Army and plagued it throughout the war, predictably emerging in the East-Prussian offensive. The 31st Tank Brigade reported being underequipped with wheeled vehicles prior to the operation. According to the table of organization, the unit required 166 trucks and truck-based specialized vehicles, but it only had 47. Fortunately, in this particular case, the deficit was covered using corps and army-level logistic battalions. Otherwise, as stated in the 31st

Tank Brigade's report, "the logistical and support units of the brigade failed to supply the advancing elements of the brigade with fuel, ammunition, and food rations."[187]

The combat experience also reaffirmed the importance of coordination and interoperability. The post-war report states that "the best results were achieved whenever fire support coordination between tanks and self-propelled guns (SPGs) was established."[188]

Steps for improving interoperability were taken at the highest level. For example, forward artillery observers were integrated into the tank formations. According to documents from the 29th Tank Corps, "artillery units assigned their best officers to tank units. These officers were stationed in tanks equipped with radio stations and tasked with requesting artillery support when tank units encountered strong resistance."[189] In addition, during the advance, artillery staffs of the tank and mechanized corps maintained direct and stable communication with the staffs of combined arms artillery units to request fire support whenever needed.

From a tactical perspective, self-propelled artillery was used in all forms of combat. In offensive operations, self-propelled guns were used for supporting tank and motor-rifle units by destroying enemy fortifications and anti-tank guns and infantry armed with man-portable anti-tank systems, such as Panzerfaust. Additionally, it served in an anti-armor role, helping to repel enemy counterattacks that were supported by tanks and self-propelled guns.

The most effective tactical formation, as was proved in combat actions, consisted of a first line of medium tanks, followed by self-propelled artillery in the second line, maintaining a distance of 400–500m between them. While two SPG batteries engaged targets from static positions, the other two batteries moved to new positions, performing a leapfrog maneuver and coordinating their actions with the tanks.

During the exploitation of the breakthrough and advance, self-propelled units were used to cover unprotected flanks. A similar task was assigned to self-propelled artillery during the pursuit of a retreating enemy. SPGs were moved to shield flanks against enemy ambushes and provide tanks with the opportunity to outflank and destroy them.

In defense, SPGs served as mobile (roving) artillery or were employed in ambushes along the tank-threatened directions. Self-propelled artillery units could be deployed *en masse*, or partially. For example, if 2–3 batteries were used as roving artillery, the remaining batteries remained in the corps commander's reserve, ready to repel enemy counterattacks from different directions.

Ambush tactics efficiently complemented the standard approach of Soviet tank and mechanized troops. Typically, these units rapidly advanced towards their designated objectives and, upon capturing them, deployed infantry and self-propelled artillery to secure captured positions. In most cases, the coordinated efforts of infantry, tanks, and self-propelled artillery resulted in the successful repulsion of enemy counterattacks. Once the positions were secured, the unit advanced toward the next objective.[190] While doctrinally, it was not recommended to position self-propelled artillery in the first line, the report of the 29th Tank Corps mentions that, under favorable circumstances, such as terrain allowing for concealed movement, self-propelled guns could move forward to counter enemy tank counterattacks. The best results were achieved when a single target was engaged by at least two self-propelled guns.

Summing up, during the operation, self-propelled artillery proved itself highly effective and was characterized in the post-war reports as "the primary means of support for tank units in all forms of modern warfare."[191] This assessment is indeed correct when discussing the outcome of large-scale operations at the strategic and operational levels. However, the performance of self-propelled artillery within the entire military system—comprising industry, materiel, doctrine, training, and organization—was more nuanced. While some elements, such as materiel and tactics, undoubtedly improved over time, others retained their defects or even degraded, as seen in the case of training and the quality of available manpower.

Among the strongest sides were a massive surge in industrial capacity and the continuous development of new models of self-propelled artillery. This factor allowed the accumulation of a sufficient mass of self-propelled artillery units by second half of 1944. Additionally, most of the models serially produced in 1944–45, while not without flaws, were generally adequate and suited the combat environment in which the Red Army operated.

Other aspects that improved in comparison with 1943 were doctrine and organizational structure. By 1945 tank and mechanized formations, combining mass, mobility and firepower, allowed the conduct of complex combined arms operations over vast areas. Masses of tanks, supported by mobile self-propelled artillery, enabled swift maneuvers and the projection of firepower to specific sectors of the battlefield as needed. The abundance of armored vehicles allowed for maintaining offensive momentum longer, despite combat losses and mechanical wear. This approach, mostly compensated the deficiencies in organization, training, as well as flaws in materiel, and broadly, in the system of armaments.

However, a major downside of this strategy was the substantial casualties in manpower and materiel. For example, before the operation that commenced on January 17, the 29th Tank Corps had 120 T-34 tanks and 112 self-propelled guns (SPGs) of all models. In just two weeks, by February 1, 1945, the 29th Tank Corps had only 15 operational tanks and 29 SPGs, losing 87.5% of its tanks and 74.1% SPGs available on January 17, 1945.[192] It is noteworthy that the high level of attrition for self-propelled guns (SPGs) was almost as significant as for tanks, regardless of their weight class or tactical role. For instance, the 31st Tank Brigade lost 60% (12 out of 20) of its heavy ISU-152 SPGs, while the 25th Tank Brigade lost 95.5% (21 out of 22). Additionally, the 32nd Tank Brigade lost 68% (17 out of 25) of its medium SU-85 SPGs.[193]

This level of attrition was not characteristic of this specific unit but was typical for Soviet mechanized formations up to the tank army and front levels, even during the late period of the Great Patriotic War. Below is Table 13 showing the losses of tanks and self-propelled guns in the formations of the 2nd Belorussian Front during the East-Prussian offensive operation between January 15 and January 27, 1945.

Even the initial examination and analysis of the available data allow us to make certain observations. First, the high ratio of tanks to self-propelled guns (1:0.7) in assault formations, like the 5th Tank Army, indicates the growing importance of self-propelled

Table 13. Combat losses suffered by tank and mechanized formations of the Belorussian Front during the East-Prussian offensive operation between January 15 and January 27, 1945.[194]

Unit	Equipment	Established Strength	Factual Strength on January 15	Under/ Overstaffed	Under/ Overstaffed, %	Total, lost	Arrived with reinforcements between 15 and 27 Jan.	Available by end of the operation	Total losses in % to Fact. Strength
5th Guards Tank Army	Tanks	305	345	40	13.1%	287	73	131	83.2%
	SPGs	210	239	29	13.8%	127	22	134	53.1%
	Total	515	584	69	13.4%	414	95	265	70.9%
8th Tank Corps	Tanks	231	211	-20	-8.7%	168	71	114	79.6%
	SPGs	42	41	-1	-2.4%	33	16	24	80.5%
	Total	273	252	-21	-7.7%	201	87	138	79.8%
8th Mechanized Corps	Tanks	196	210	14	7.1%	249	111	72	118.6%
	SPGs	42	42	0	0.0%	16	5	31	38.1%
	Total	238	252	14	5.9%	265	116	103	105.2%
1st Guards Tank Corps	Tanks	210	51	-159	-75.7%	74	52	29	145.1%
	SPGs	63	166	103	163.5%	171	66	61	103.0%
	Total	273	217	-56	-20.5%	245	118	90	112.9%
Total, Tanks		942	817	-125	-13.3%	778	307	346	95.2%
Total, SPGs		357	488	131	36.7%	347	109	250	71.1%
Grand Total		1299	1305	6	0.5%	1125	416	596	86.2%

artillery. Since its almost complete nonexistence in 1941, and through its introduction in 1943, this new arm of service had evolved into a highly valuable operational-level asset that has found its rightful place in the Soviet Tank and Mechanized Force.

Note that a similar ratio of 1:0.6, or 3 self-propelled guns to 5 tanks, is observed across the mechanized units of the entire 2nd Belorussian Front. The emphasis on self-propelled artillery is further supported by the fact that some formations, such as the 1st Guards Tank Corps, were understaffed with tanks but instead equipped with self-propelled guns.

The second observation is the equally high level of attrition for both tanks and self-propelled guns. This suggests that, despite all efforts and improvements, the employment of self-propelled artillery was far from ideal even in 1945. Sometimes it was employed contrary to the doctrine and often not properly supported, which, as a result, is reflected in the figures of losses.

In other words, while the Soviet operations during the late period of the Great Patriotic War were successful, and self-propelled artillery contributed greatly to it, this achievement often came at a high cost in lives and materiel. The entire system in which self-propelled artillery operated still faced challenges and, in fact, was incomplete. In 1945, it marked the very beginning of the long journey towards the "artillery army with a lot of tanks," as formulated by Lester W. Grau and Charles Bartles.[195]

Chapter 6

Allied Assistance

As was discussed in the previous chapters, by 1942, the Soviet leadership recognized the need for self-propelled artillery and initiated a series of research and development projects. While the development of direct fire support self-propelled guns moved at an acceptable pace, work on other classes of self-propelled artillery went slower than expected and encountered considerable difficulties. The most problematic of them, unsurprisingly, were anti-aircraft and indirect fire self-propelled guns, as well as self-propelled artillery on wheeled and half-tracked platforms.

For example, since 1942, the Soviet military leadership was extremely concerned about the threat posed by enemy attack aircraft and was seeking ways to improve anti-aircraft capabilities for tank units. On April 15, 1942, the plenary session of the Artillery Committee of the Main Artillery directorate made the decision to start development of self-propelled anti-aircraft guns.[196] While the first draft designs were already introduced in the spring of 1942 and the first prototypes were delivered by December of the same year, these experimental vehicles were mostly unsuccessful and failed the initial trials.

As a result, the Soviet leadership was forced to investigate other possibilities, such as ordering these vehicles through Lend-Lease. The first Allied self-propelled guns that were ordered and arrived in the USSR were US-made T48 57mm Gun Motor Carriages. Initially they were designed for the British army in the Northern Africa, however by the time these vehicles were delivered, the campaign in Africa was over and 650 vehicles were transferred to the Soviet Union.[197] According to available Soviet documentation, 241 T48 SPGs were delivered in late 1943, and the remaining 409 vehicles in 1944.[198]

Interestingly enough, the T48 SPG was tested in March 1943 at the NIBT proving grounds, and the commission recommended not adopting the vehicle.[199] The report stated that the US-made T48 SPG "has poor speed while moving on country roads or through snow-covered plain." Additionally, off-road performance and obstacle-crossing capability were deemed unsatisfactory. Among other drawbacks, the commission mentioned inadequate operating range and unreliable transmission.[200] Despite that, the T48 SPG was adopted into service with the Red Army under the designation SU-57 or sometimes SU-57-I ("I" stands for "inostrannii" or "foreign").[201]

The M15 and M17 SPAAGs, along with the M10 and T70 tracked SPGs, were arriving in the Soviet Union throughout 1944. Deliveries of the M15 and M17 began in May 1944, while the M10 and T70 arrived in January and February.[202]

The M10 SPG was tested in September–October 1943 and received generally positive feedback. The testing commission noted good mobility and maneuverability of the vehicle, as well as positively evaluated turret design and working conditions for the crew, including the driver. Another advantage of the American SPG was its commonality with M4A2 medium tanks and its "simple in design suspension system ensuring smooth movement."[203] According to the commission, two major drawbacks of the M10 were unreliable running gear, in particular the road wheels, and the design of the open-topped turret.[204] Both complaints, however, seem to be a bit of a stretch, as faulty running gear and an open-topped fighting compartment were two known and widely criticized drawbacks of the most common Soviet SPG, the SU-76. Eventually, the M10 SPGs were ordered in the US. The first and only batch, however, was limited to 52 units. The M10 SPGs in Russian service were known as SU-76 or SU M-10.

Another tracked SPG, the Gun Motor Carriage T70, faced a completely different fate. The SPG was tested from March 20 to May 8, 1944 and was not recommended for import. The commission stated that the T70 had limited off-road capability, poor stability even on paved roads and weak protection. However, the most concerning issue was the excessive fuel consumption. According to the test results, the T70 required 78.5 liters of fuel per hour even while moving on paved roads and 80.5 liters on country roads, whereas the fuel consumption exceeded 130 liters per hour when crossing snow-covered fields.[205] While some of the issues were pointed out correctly, it is worth noting that the USSR received a batch of pre-production T70 prototypes that were improved later.

On the other hand, the arguments for ordering an additional fire-support vehicle on a tracked platform were not justified enough for the military leadership. By 1944, the Soviet industry had provided the army with domestically manufactured self-propelled artillery on tracked chassis, including the light SU-76 armed with a ZiS-3 76mm gun and medium SU-122 and SU-85. In addition, the Soviet Union was already importing M4A2 medium tanks. Armed with practically the same gun as the T70 and considerably better protected, these vehicles were often used in the direct fire support role or as self-propelled artillery.

Probably the most successful vehicles among the imported SPGs were the M15 and M17 self-propelled anti-aircraft weapons. These vehicles began arriving in the Soviet Union in May 1944. In early September 1944 both vehicles were tested at the NIBT proving grounds and recommended for import and service with the Red Army.[206] Surprisingly, no significant drawbacks were identified for both vehicles. On the contrary, the trials commission characterized them as "reliable and effective."

Even more praise was given for the M15 and M17 by Soviet troops. Both vehicles were valued so highly that the Commander of the Tank and Mechanized Forces of the Red Army, Yakov Fedorenko, approached Anastas Mikoyan, Deputy Chairman of the Council of People's Commissars (SNK) and Council of Ministers of the USSR, with a request to "procure 2000 units of each type of these self-propelled guns."[207]

It is worth noting that in 1944 and early 1945, the end of the war was not yet in sight, so the Soviet military leadership was preparing procurement plans for the near future. For example, the Military Council of the Tank and Mechanized Forces (BT and MV) wanted to order as many as 5,000 M4A2 tanks for campaigns in 1945–46.[208]

Table 14. Deliveries of US self-propelled artillery through Lend-Lease.[209]

Year	M10	T70	M15	M17	M48
1943	–	–	–	–	241
1944	52	5	100	1000	409
1945	–	–	–	–	–

From a tactical employment perspective, the self-propelled artillery of foreign origin had different experiences, service histories, and impacts on the development of the Soviet mechanized forces during and even after World War II.

The M10 self-propelled guns were issued to the 1239th and 1223rd Light Self-Propelled Regiments (LSAP). In most instances, these vehicles were employed in the same way as domestically produced SU-76M SPGs. For example, the M10 SPGs of the 1223rd LSAP provided direct fire support for infantry units and the 31th Tank Brigade.

There are, however, documented episodes of M10s being used as tanks. The reason for this, according to the combat report of the 1223rd LSAP, was that the tank unit it supported had suffered heavy losses and there were more SPGs available at that moment. The SPGs were ordered to attack enemy positions, but they were "attacked with hand-grenades by enemy infantry while passing through the first line of enemy defenses." As a result, three M10 SPGs were damaged.[210] Generally, the command of the 1223rd LSAP was satisfied with the performance of the M10 and highlighted such advantages as mobility along with quality and precision of the gun optics. The main disadvantage was the height of the vehicle, which, according to the combat report, "made it a large-sized target." Other negative issues included unreliable elements of the running gear, such as clutch, tracks and roadwheels.[211]

The vast majority of T48 (SU-57) self-propelled guns went to light brigades of self-propelled artillery, or SABR. There were three brigades equipped with T48 self-propelled guns: 16th, 19th, and 22nd SABR. These brigades were formed according to TO&E #010/508 and comprised of 63 self-propelled guns in three battalions (divisions) of 21 SPG, plus a command company of 5 SPGs.[212] The SU-57 SPGs were also issued to reconnaissance battalions and companies, as well as to separate motorcycle regiments and battalions. Initially, the brigades armed with T48 SPGs were meant to act as typical light self-propelled artillery brigades armed with SU-76M SPGs. However, over time, the SU-57s went far beyond the fire-support role. The 16th SABR, for example, was a mobile reserve of the Tank Army's commander and employed for screening the open flanks of the Tank Army advancing in the enemy's operative depth. In fact, the SU-57 SPGs were used in the same role that the French AMX-10 heavy reconnaissance vehicles were designed for. Another role of the brigade was deep raids to the enemy's rear. The Soviet command appreciated the advantages of SU-57 SPGs that combined the firepower of the 57mm gun with the speed and mobility of the M3 half-track chassis.

The 16th Brigade combat report mentions that "the operations of the brigade were swift, bold, and decisive. The swift maneuver while operating in the enemy's depth fully justified itself. The brigade engaged the enemy where they least expected it. As a result, the brigade suffered only minor losses in men and materiel and at the same time inflicted heavy losses on the enemy."[213] Additionally, the brigade was used for capturing and securing key areas (towns, villages, and airfields) and as a part of the advance guard force of the Tank Army.[214]

The commander of the 16th SABR went even further and proposed to bolster the brigade with "rocket artillery for indirect fire, a battalion of motorized infantry, and mobile anti-air means." This formation, in his opinion, would be perfectly suited for operations in the enemy's rear after the breakthrough is made. The author believed that such a brigade with force multipliers would be able to perform as a mobile anti-tank reserve as well as secure key bridges and river crossings or engage enemy reserves in the rear.[215] In some sense, the author foreshadowed such a modern concept as Medium Weight brigades (or Strike brigades for the British army) that offer increased mobility along with high firepower and autonomy.

M15 and M17 SPAAGs were initially intended as mobile means of air defense for fast-moving armored formations. Feedback provided by army units for M15 and M17 was incredibly positive.

> Imported American self-propelled guns, such as M15 and M17, served their intended purpose. Due to the fact that self-propelled guns are mounted on maneuverable M2, M5 and M9 armored vehicles with good cross-country capability, they are used to accompany armored forces as means of protection against enemy aviation. It should be noted that the strong concentrated fire from the M15, which features a 37mm cannon paired with two large-caliber anti-aircraft machine guns, is an effective weapon against enemy aircraft. The same can be said for the M17 self-propelled gun, which has a quadruple mount of large-caliber anti-aircraft machine guns and releases bursts of bullets towards enemy planes.[216]

However, the M15 and M17 also proved themselves versatile and capable in other combat roles. In the 6th Guards Motor-Rifle Division, the M17 SPGs were used together with M3A1 APCs for reconnaissance. According to the report, these vehicles were used for both mounted reconnaissance and for providing fire support to dismounted scouts.[217] In addition, the M17s were used for "limited counterattacks" and sometimes for transporting troops.[218]

The documents of the 45th Guards Tank Regiment of the 9th Guards Tank Brigade give more insight on the combat employment of the M17 MGMC. Firstly, M17 platoons were used for air defense in "all kinds of combat operations." The first platoon was attached to the advance tank battalion, while the second was attached to the tank battalion in the second echelon, or reserve tank battalion. The third platoon was near the command post of the brigade. The M17s proved themselves as a 'strong air defense weapon, provided that the SPGs are employed in groups of 3–6 vehicles." According to the document, in only two weeks (from April 15 to April 30, 1945), the brigade's anti-aircraft company destroyed 6 German planes using M17 SPAAGs.[219]

Another typical job for the M17 SPAAGs was supporting tank units against ground targets in difficult environments, such as forests or urban terrain. In the Battle for Berlin, the M17 SPGs were used with the 45th Guards Tank Regiment in urban warfare as part of assault detachments, accompanying tanks, artillery, and infantry. Using their "strong machine gun fire, they suppressed or destroyed positions of enemy infantry and anti-tank detachments" armed with Panzerfaust and Panzerschreck rocket launchers. When advancing behind tanks through the streets of the city, they fired at windows, attics, and entryways, suppressing the enemy and preventing them from returning fire. On some occasions, the M-17 APCs were used to set buildings on fire."[220]

To some extent, the combat employment of the M17 foreshadowed the use of Soviet SPAAGs in urban warfare during the Cold War and in modern conflicts. Ultimately, it paved the way for the concept that materialized in a new class of heavy armored combat vehicles, such as the BMPT. Finally, during the operations in the enemy operative depth, the groups of 2–3 M17 were used for route reconnaissance and "demonstrated good results." The documents of the 51th Separate Motorcycle Regiment mention "dug-in" M3A1 APCs and M17 SPAAGS used in static defense.[221]

Both M17 and M15 vehicles were often used as mobile fire support platforms, thanks to their superior firepower. The report of the 48th Army under the title "The Combat Use of the Armored Personnel Carriers and Armored Cars in the Great Patriotic War" describes a combat episode when a group of M3A1 APCs accompanied by M15 SPAAGs attacked a German infantry in the town of Langwalde (today Długobór in Poland). As described in a report, "the fire was so dense that the tracer rounds seemed to touch each other in the air." As a result, the German infantry was overwhelmed and began retreating in disorder, and the town was captured by the Soviet troops shortly after.[222] Summing up, the M15 and M17 were used in all possible combat scenarios and roles: in offense and defense, against air and ground targets, for air defense, and as mobile fire support platforms, as well as for screening, reconnaissance, and pursuit of retreating enemy forces.

Surprisingly, the Soviet leadership and procurement committee showed little interest in indirect-fire self-propelled artillery. An explanation for this may be that the Soviet military leadership simply did not see the need to procure this type of equipment. Theoretically, Soviet self-propelled artillery was capable of firing indirectly, although in practice, this was rarely practiced. Another explanation may be a lack of understanding at the doctrinal level and the low qualifications of the Soviet military experts responsible for procurement.

This version is supported by documented criticism coming from the Engineering Directorate of the People's Commissariat of Foreign Trade (NKVT). The NKVT, for example, complained that "The attitude of the Armored Directorate towards imported equipment leaves much to be desired. Responsible personnel of the Tank and Mechanized Forces have difficulties determining which vehicles they require, which have gained recognition, and which ones should no longer be procured."[223]

The NKVT provides an example of incompetency:

We imported over 2000 Mk-1 armored personnel carriers and approximately 3000 M3A1 armored vehicles. These machines are actively engaged in combat in the armored units of the Red Army. Over a period of 2.5 years, we have been trying to gather any feedback on these machines, but we have not been able to obtain a definitive response. At one point, they are needed, and at another moment, they are not. At one point, they are considered good, and at another, nobody seems to know anything about them. If they are not imported, the Armored Directorate asks when and how many will be delivered because they urgently need them. However, if they are being imported, they complain and question why these machines are being supplied to them. Due to such an attitude from the Armored Forces Command towards the machines, we had to refuse further procurement from them.[224]

Table 15. Orders and delivered materiel from October 1, 1941 to September 15, 1945, according to data provided by Soviet commission in the US.[225]

Designation (Soviet)	Designation (US)	Requested	Delivered to USSR
M10 Tank Destroyers	M10	2	2
Guns 3in anti-tank, self-propelled	M10	50	50
Guns 37mm anti-air, self-propelled	M15 Combination Gun Motor Carriage	100	100
Quad machine guns 50 cal	M17 Multiple Gun Motor Carriage, M17 MGMC	1000	1000
Guns anti-tank, 57mm, self-propelled (SU-57)	T48 57mm Gun Motor Carriage	650	650
Guns anti-tank, 76mm, self-propelled	GMC T70	5	5

Table 16. Foreign self-propelled artillery delivered to USSR on April 1, 1945.[226]

Designation (Soviet)	Delivered on January 1, 1945	Available on April 1, 1945	In the Army	In Military Districts	At Repair Facilities	Irrecoverable losses on January 1, 1945
M10 Tank Destroyers	52	47	14	7	26	5
SU-57 (T-48)	650	512	336	90	86	111
T-70	5	5	0	5	0	0
M-17	1000	1000	786	214	0	0
M-15	100	100	87	13	0	0

In total, by the end of the Great Patriotic War the Soviet Union received 1807 US-made self-propelled guns. The vast majority of them (96.8%) were half-track based M15, M17 and T48 combat vehicles.

While self-propelled artillery provided through Lend-Lease represented only a small fraction of the Soviet fleet of self-propelled guns, the role of these vehicles should not be underestimated for several reasons. Firstly, these vehicles were invaluable in closing critical capability gaps such as anti-aircraft defense for fast-moving armored formations or mobile means of direct fire support for reconnaissance and assault units. Secondly, the Soviet engineers got an opportunity to examine foreign technology in detail. In some instances, they recommended adopting specific technical solutions or components. For example, deputy head of the NIBT proving grounds Engineer-Lieutenant Colonel A. M. Sych wrote in the post-trial report of the M17 MGMC in September 1944: "The design of the M45 machine-gun mount could serve as an example for designing anti-air mounts by domestic industry."[227] Finally, the employment of self-propelled guns imported from the United States has provided the Soviet military with invaluable combat experience and knowledge. This experience, generalized and analyzed, has played its own important role in the post-war development of the Soviet armored forces.

It is worth emphasizing that the wartime experience was not conceptualized and implemented immediately in the aftermath of the Great Patriotic War. As a result, the Soviet Army received certain types of self-propelled artillery later than expected, while the others were underrated by the Soviet military and never adopted.

For example, domestically produced wheeled APCs and SPAAGs were adopted by the Soviet Army only in the early 1950s, while the tracked SU-57-2 SPAAG was adopted in 1955. Before that, the Soviet Army had only 75 armored SPAAGs that were able to accompany tank formations and provide cover against enemy aircraft. Considering that aviation had already entered the era of jet-powered aircraft, the Soviet Army was at a severe disadvantage.

Conclusion

The Perfect Combination

Since the Middle Ages, the Russian army has heavily relied on artillery. With new technical capabilities emerging in the Industrial Age, it was only a matter of time before military engineers would conceive the idea of combining artillery firepower with the mobility offered by automobile or tractor chassis.

The first self-propelled guns were introduced and adopted by the Russian Imperial Army in the early months of the Great War. Since then, this new arm of service enjoyed a stable evolution, uninterrupted even by the dramatic events that followed the Russian Revolution. The introduction of the Deep Battle doctrine and the program of mechanization and motorization of the Red Army in the 1930s marked a significant leap forward in the development of self-propelled artillery. During this period, the theoretical basis was formed, and all main classes of self-propelled guns were developed, with some of them tested in the local conflicts of the 1930s and put into low-rate serial production. Remarkably, there was a strong emphasis on indirect fire self-propelled guns and self-propelled anti-aircraft guns.

The rapid development of the Soviet mechanized forces was abruptly disrupted as a result of Stalin's Great Purge. Many theoreticians, high-level commanders, and engineers were brought to trial, subsequently accused of wrecking and treason, and either executed or removed from their positions and imprisoned. The development and production programs of armored vehicles, including self-propelled artillery, also suffered greatly, with many of them being deprioritized and canceled.

These events effectively set back the evolution of Russian self-propelled artillery by decades, both technically and theoretically. Consequently, the Red Army entered the Great Patriotic War with only a handful of prototypes developed during the 1930s that had survived either at testing sites or in army warehouses.

The fierce battles of 1941–42 revealed weaknesses in the Soviet armored doctrine and gaps in the armaments system. The Soviet leadership struggled to fill the capability gap with low-rate production and makeshift vehicles. By the spring of 1942, the acute need for self-propelled artillery was obvious. In October 1942, the State Defense Committee initiated the construction of prototypes for a 122mm self-propelled howitzer, a 45mm anti-tank gun, and a 76mm assault gun. By the end of 1942, self-propelled artillery was reintroduced as an arm of service.

After the initial employment in early 1943 and the first employment *en masse* during the Battle of Kursk, self-propelled artillery entered a period of continuous growth and improvement. The Soviet military received the opportunity to gain and accumulate battlefield experience, which was used to enhance theory, combat manuals, regulations, and improve the design of combat vehicles.

be underestimated for several reasons. Firstly, these vehicles were invaluable in closing critical capability gaps such as anti-aircraft defense for fast-moving armored formations or mobile means of direct fire support for reconnaissance and assault units. Secondly, the Soviet engineers got an opportunity to examine foreign technology in detail. In some instances, they recommended adopting specific technical solutions or components. For example, deputy head of the NIBT proving grounds Engineer-Lieutenant Colonel A. M. Sych wrote in the post-trial report of the M17 MGMC in September 1944: "The design of the M45 machine-gun mount could serve as an example for designing anti-air mounts by domestic industry."[227] Finally, the employment of self-propelled guns imported from the United States has provided the Soviet military with invaluable combat experience and knowledge. This experience, generalized and analyzed, has played its own important role in the post-war development of the Soviet armored forces.

It is worth emphasizing that the wartime experience was not conceptualized and implemented immediately in the aftermath of the Great Patriotic War. As a result, the Soviet Army received certain types of self-propelled artillery later than expected, while the others were underrated by the Soviet military and never adopted.

For example, domestically produced wheeled APCs and SPAAGs were adopted by the Soviet Army only in the early 1950s, while the tracked SU-57-2 SPAAG was adopted in 1955. Before that, the Soviet Army had only 75 armored SPAAGs that were able to accompany tank formations and provide cover against enemy aircraft. Considering that aviation had already entered the era of jet-powered aircraft, the Soviet Army was at a severe disadvantage.

Conclusion

The Perfect Combination

Since the Middle Ages, the Russian army has heavily relied on artillery. With new technical capabilities emerging in the Industrial Age, it was only a matter of time before military engineers would conceive the idea of combining artillery firepower with the mobility offered by automobile or tractor chassis.

The first self-propelled guns were introduced and adopted by the Russian Imperial Army in the early months of the Great War. Since then, this new arm of service enjoyed a stable evolution, uninterrupted even by the dramatic events that followed the Russian Revolution. The introduction of the Deep Battle doctrine and the program of mechanization and motorization of the Red Army in the 1930s marked a significant leap forward in the development of self-propelled artillery. During this period, the theoretical basis was formed, and all main classes of self-propelled guns were developed, with some of them tested in the local conflicts of the 1930s and put into low-rate serial production. Remarkably, there was a strong emphasis on indirect fire self-propelled guns and self-propelled anti-aircraft guns.

The rapid development of the Soviet mechanized forces was abruptly disrupted as a result of Stalin's Great Purge. Many theoreticians, high-level commanders, and engineers were brought to trial, subsequently accused of wrecking and treason, and either executed or removed from their positions and imprisoned. The development and production programs of armored vehicles, including self-propelled artillery, also suffered greatly, with many of them being deprioritized and canceled.

These events effectively set back the evolution of Russian self-propelled artillery by decades, both technically and theoretically. Consequently, the Red Army entered the Great Patriotic War with only a handful of prototypes developed during the 1930s that had survived either at testing sites or in army warehouses.

The fierce battles of 1941–42 revealed weaknesses in the Soviet armored doctrine and gaps in the armaments system. The Soviet leadership struggled to fill the capability gap with low-rate production and makeshift vehicles. By the spring of 1942, the acute need for self-propelled artillery was obvious. In October 1942, the State Defense Committee initiated the construction of prototypes for a 122mm self-propelled howitzer, a 45mm anti-tank gun, and a 76mm assault gun. By the end of 1942, self-propelled artillery was reintroduced as an arm of service.

After the initial employment in early 1943 and the first employment *en masse* during the Battle of Kursk, self-propelled artillery entered a period of continuous growth and improvement. The Soviet military received the opportunity to gain and accumulate battlefield experience, which was used to enhance theory, combat manuals, regulations, and improve the design of combat vehicles.

The evolution of self-propelled artillery can only be assessed within the broader system that includes such subsystems as industry, organizational structure, training, and doctrine. Throughout the Great Patriotic War, these elements underwent changes but evolved unevenly, leading to varied outcomes.

Of all the elements in the system, production and design were likely the most successful. Despite losing control over a significant part of the territory in 1941–42 and the subsequent evacuation, Soviet military industry was able to adapt and significantly increase the output of self-propelled artillery. While in 1941–42 it produced only 50–100 self-propelled guns, 1943 saw an increase to over 4000 SPGs, with the peak production reaching 12,060 SPGs in 1944 and over 6000 SPGs in the first half of 1945. In total, between the beginning of the Great Patriotic War and the capitulation of Japan in September 1945, as many as 24,814 self-propelled guns were produced by the Soviet industry.[228] Other sources provide slightly lower figures. For example, Melnikov cites 22,436 SPGs produced during the war,[229] while Ermolov gives a total of 22,411 SPGs,[230] which could be explained by different methods of calculation. One of the crucial prerequisites that enabled high production output was the construction of simplified designs around existing systems (guns, engines) and serially produced vehicles, such as IS, KV-1, T-34, or T-70 tanks. The extensive production, in turn, enabled the rapid growth of tank and mechanized forces in general and self-propelled artillery in particular.

However, there were substantial drawbacks. At the lower levels, the continuous emphasis on quantity led to numerous issues with quality and persistent design flaws. It's worth noting that some drawbacks arose as a consequence of decisions aimed at overcoming a crisis but remained in place even as the situation improved. A vivid example of this is the decision made in December 1941 to exclude grinding and mechanical processing of armor and welded seams from the technological cycle to simplify and streamline production. From that time onwards, tanks and self-propelled guns were leaving factories "with a rough, unpolished surface of the armor and thick welded seams, which could sometimes even cause injuries."[231]

Although some of the problems were gradually resolved, others persisted until the end of the war. These issues included the mechanical reliability of the running gear, especially with gearboxes, tracks, and ball bearings, the quality of optics and radio equipment, as well as a broad spectrum of problems related to situational awareness. Another set of issues, consistently running through testing and combat reports like a common thread, was related to ergonomics.

At the higher level, the industry was reluctant to change well-established production cycles that guaranteed high production output and, consequently, the fulfillment of state plans. For instance, the industry prioritized the production of SU-76 self-propelled guns, which, according to the military, were largely inadequate for warfare in late 1944 and 1945. SU-76 light self-propelled guns made up 55–56% of the total production of Soviet self-propelled guns.

Consequently, the Soviet industry failed to address some critical capability gaps that negatively affected the performance of Soviet tank and mechanized troops. For example, during the Great Patriotic War, the Soviet military did not receive indirect fire self-propelled guns, mass-produced self-propelled anti-aircraft guns, or armored ammunition transporters crucially needed for supplying tanks and SPGs, as well as self-propelled artillery on wheeled or half-track platforms. Some of these gaps, however, were partially filled with the help of the Lend-Lease agreement.

Another, at least partially, successful aspect of the system was the organization and structure of the tank and mechanized forces and self-propelled artillery units. The Soviet military and political leadership succeeded in establishing a large and mostly effective organizational structure and a system of command, training, and supply for self-propelled artillery. In addition, self-propelled artillery as an arm of service was organized at a remarkably rapid pace.

The positive aspect of this was the quick adoption and integration of self-propelled artillery into the system of tank and mechanized troops. On the other hand, some decisions were of a temporary nature or made hastily, requiring further improvements and reorganizations. Consequently, the organizational structure of tank and mechanized forces, including self-propelled artillery, was the subject of continuous evolution throughout the entire Great Patriotic War. These frequent changes created confusion among the troops, leading to improper or ineffective use of self-propelled artillery.

In essence, the Soviet military and industry succeeded in developing self-propelled guns that were adequate for addressing the acute needs of the Red Army and producing them in sufficient numbers to saturate troops with self-propelled artillery. Equally successful was the establishment of a system capable of reproducing self-propelled artillery units at a remarkably high pace. While both aspects retained some significant drawbacks, they ensured numerical superiority for the Red Army over the Wehrmacht in self-propelled artillery. However, as with materiel production, the negative aspect of this "factory of units" was their quality.

The availability and quality of manpower had the most substantial negative influence over the entire system. This problem originated in the early 1930s when the army began its transformation towards mechanization and motorization and, on a larger scale, was a characteristic challenge for the entire Red Army. By 1943, the problem had significantly worsened, following the explosive growth of Soviet tank and mechanized forces, which now included self-propelled artillery. As a result, the Soviet Tank and Mechanized Forces, and more broadly, the Red Army, faced two opposing trends: on the one hand, the army was in constant need of qualified and high-quality manpower. On the other hand, the reserves of qualified manpower were depleted by the Great Purge, the intensive growth of the army, and the need to replenish losses suffered in 1941–42. The Main Directorate of Troop Formation and Personnel Assignment of the Red Army (UFiU RKKA) highlighted a negative trend in the overall quality of manpower assigned to tank, mechanized, or self-propelled units. According to post-war reports, the quality of recruits throughout the Great Patriotic War was evaluated as "satisfactory." In 1941, 1942, and partially in 1943, the incoming reinforcements were assessed as "good in quality and well-trained." However, from the end of 1943 onward, the quality of reinforcements and their training started to deteriorate. The same negative trend persisted in 1944 and 1945.[232] The UFiU highlighted that "the crews of tanks and self-propelled guns were inadequately trained. The vast majority of drivers had practical experience between 10 and 25 hours."

The same problems applied to other military personnel allocated to tank and self-propelled artillery units. Recruits born between 1926 and 1927 and assigned as truck drivers were trained so poorly that using them as drivers without any additional training was considered dangerous and was recommended to be prohibited. Similar issues were noted among personnel assigned to submachine-gunner and sapper units of the tank and self-propelled artillery units and formations.

"The training of these troops is inadequate—only one to three months at reserve regiments—and it did not meet the requirements of the Tank and Mechanized Forces. The sappers have the lowest level of training, both in their professional field of knowledge and in the specifics of tank units" operations, while the riflemen were never trained to accompany tanks as tank riders (tankodesant)," stated the UFiU report.[233]

The physical fitness of recruits was another frequently mentioned issue. The UFiU noted that in 1943, 1944, and 1945, the reserve rifle regiments often sent reinforcements to tank and mechanized units with "STDs, open wounds, limited mobility, and other impairments," making them unfit for service in fast-moving troops. Finally, there were servicemen without any or very poor Russian language skills, mostly recruited in Soviet Socialist republics in Middle Asia (Kyrgyz SSR, Kazakh SSR, and Tajik SSR) or in the Soviet republics of Transcaucasia.

The quality and availability of manpower were crucial factors affecting both individual and collective training for self-propelled artillery units. Fundamentally, the issue stemmed from initially under-skilled soldiers and officers who underwent accelerated training in the rear. As a result, the training system produced underprepared troops that needed partial or complete retraining within frontline units. This, in turn, meant that a significant portion of the efforts made by the training system in the rear were fruitless and only wasted time and scarce resources.

While some of these issues were addressed, the solutions were rarely permanent and maintained their own shortcomings. For example, to resolve the problem of poor tank driver training, the Commander of the Tank and Mechanized Forces of the Red Army, Yakov Fedorenko, allowed the frontline units to replace the unskilled personnel with their own battle-hardened and experienced drivers.[234] On a broader scale, the challenges in collective training inevitably affected the ability of self-propelled artillery units to interact seamlessly with other branches of service, units, and elements within combat formations.

As formulated in the US Army's doctrine, the essence of combined arms warfare lies in the synchronized and simultaneous application of different weapons, aiming to achieve a greater effect than if each arm were used separately or sequentially.[235] This level of interoperability was often lacking. Typically, this led to degraded performance on the battlefield, frequently resulting in casualties or failed missions.

Two notable examples from the documents of the 31st Tank Brigade of the 29th Tank Corps and from the inspection report of the self-propelled artillery regiments of the Western

Front vividly illustrate the consequences when one or more elements in a combined arms battle were not in the right place or did not function properly.

The first report of the 31st Tank Brigade mentions that during the East-Prussian operation, the unit suffered severe losses in tanks and self-propelled guns. Between February 14 and 24, 1945, the unit lost all of its T-34 tanks and ISU-152 heavy self-propelled guns (8 tanks and 8 SPGs lost, 9 irrecoverably). These losses were attributed to the inappropriate use of tank units in offensives against heavily fortified enemy positions without sufficient infantry support. Other contributing factors to substantial tank losses included a lack of reconnaissance and poor skills of accompanying towed artillery. As described in the report, "the artillery conducted fire in most cases to empty areas, and poorly adjusted fire during the preparation."[236]

The second report mentions an interesting German tactic employed against Soviet self-propelled guns at the Western Front in September–October 1943. According to the report, German gunners "literally hunted for our self-propelled guns in every tank attack, aiming to damage the guns or gun mantlets." This tactic proved highly effective, as Soviet units, lacking repair and recovery capabilities, had to send the damaged self-propelled guns to factories, effectively taking them out of the battlefield for months. The report emphasizes that "the entire Western Front does not possess sufficient capabilities for repairing damaged armaments of self-propelled guns."[237] It should be emphasized that cases like those mentioned here were not uncommon and are frequently reported in the combat records and war diaries of Soviet tank and mechanized units.

The final element of the system to be examined is doctrine and tactics. Similar to organization, doctrine underwent continuous changes throughout the entire Great Patriotic War. This was partly in response to the evolving nature of warfare and the advancements in materiel. However, another significant portion of changes occurred because the early theoretical assumptions were erroneous.

In essence, the evolution of the doctrine during the Great Patriotic War reflected the difference between the initial tactical or technical ideas embedded in the designs of certain Soviet self-propelled guns, and the real combat experience, which often radically contradicted the early theoretical expectations of how self-propelled artillery was supposed to act in combat.

A striking difference exists between the doctrinal documents issued between 1942 and the first half of 1944 and the tactical recommendations developed and issued at the front, army, and divisional levels from the second half of 1944 onwards. This observation similarly applies to the tactical-technical requirements for self-propelled artillery formulated between 1942 and late 1944, as well as to the actual use of self-propelled artillery in combat, whether deployed independently or as part of combined arms formations.

Another aspect was the ability of the tank or mechanized forces to employ self-propelled artillery in accordance with doctrine and perform at a level sufficient to meet the requirements outlined in the doctrinal documents. Many combat reports emphasized the correct use, following the provided instructions and manuals, as a major prerequisite for success in combat.[238] On the contrary, whenever self-propelled artillery was used in contradiction to the doctrine it suffered catastrophic losses and was not able to achieve the assigned tasks.

In reality, the employment of self-propelled artillery and its interaction with other branches of service depended more on the individual skills of the tactical commander or the situation on the battlefield than on the specific features of the class or intended function. Heavy self-propelled guns were frequently used in an anti-tank role, and tank destroyers for direct infantry support; broadly, SPGs were employed in a variety of combat and auxiliary roles, including APCs, ammunition carriers, reconnaissance vehicles, and others.

One of the most illustrative and widespread problems related to the employment of self-propelled artillery was its use as tanks. In 1943, when self-propelled artillery was introduced to the army, the reasons for improper use, at least partially, stemmed from a lack of expertise. However, this issue persisted throughout the entire Great Patriotic War, despite significant improvements in organizational structure, materiel, and the accumulation of combat experience.

The core of the problem lay in the constraints posed by the actual capabilities of the army, including the level of command, training, interoperability, and others. These constraints largely determined the form of employment of self-propelled artillery, as well as its configuration.

This resulted in two consequences. Firstly, self-propelled artillery as an arm of service was employed in a simplified form, with self-propelled guns of all classes mainly used for close-range direct fire support. This practice was so widespread across the tank and mechanized force that it found reflection even at the doctrinal level. For example, an examination of combat manuals and recommendations from the late period of the Great Patriotic War reveals that there were no significant differences in the tactical roles and employment of heavy tank regiments armed with IS tanks and heavy self-propelled artillery regiments. Additionally, tanks were employed in self-propelled artillery roles as frequently as self-propelled guns were used as tanks.

Secondly, some forms of employment and classes of combat vehicles were entirely omitted. For example, the requirements for the personnel needed for indirect fire self-propelled artillery were likely the main reason why Soviet self-propelled artillery never evolved beyond the direct fire support role. Given the problems that the system faced in the basic training of military personnel for self-propelled artillery units, the task of training personnel possessing the knowledge and skills of both tankers and artillerymen in a short period and at a sufficiently high level appears practically unattainable.

Some scholars and researchers believe that the Great Patriotic War provided the impetus for the development of the Soviet tank and mechanized forces, with the peak of evolution in doctrine and materiel reached in 1944–45. According to the other commonly accepted narrative, the performance of the Soviet tank and mechanized forces during the war was highly effective. The accumulated combat experience laid the foundation for further evolution, ultimately resulting in the emergence of the formidable Soviet armored force of the late Cold War.

In reality, the development of Soviet tank and mechanized forces, including self-propelled artillery, was more nuanced. Firstly, the evolution of Russian and later Soviet self-propelled artillery was uneven, with certain elements of the system experiencing varying degrees of advancement over time. The dramatic events of the 1930s set back the development of Soviet armored forces by almost a decade in terms of theory and equipment. This setback resulted in entering World War II without self-propelled artillery being a fully-fledged arm of service, despite the substantial achievements made during the 1930s. While the evolution of Soviet self-propelled artillery between 1942 and 1945 may appear as a leap forward, considering the circumstances, it was, in fact, a slow and painful recovery process toward the pre-war level.

Secondly, after its reintroduction in 1943, self-propelled artillery was warmly received and appreciated by the army. Since then it experienced steady qualitative and quantitative growth, with a trend toward expanding its combat capabilities. The significance and value of this asset were understood at the highest levels of command, leading to the introduction of higher-level units in 1944–45. The trend, where self-propelled artillery units evolved into force multipliers for tank and combined arms armies, is also highly indicative.

Thirdly, despite a period of neglect, self-propelled artillery found its rightful place in the system of armaments of the Tank and Mechanized Forces. However, the system and its elements were evolving incoherently. On one hand, the Soviet military and industry invested significant resources in the numerical growth and development of newer, more reliable, and more efficient equipment. On the other hand, this sophisticated equipment, produced under challenging wartime circumstances, was often operated by undertrained personnel and used erroneously.

On a larger scale, incoherence in one part of the Red Army's structure often led to drawbacks in others, a characteristic also observed in the system of self-propelled artillery. For example, despite the strong emphasis placed on the development of self-propelled indirect fire and anti-aircraft guns in the 1930s, these combat vehicles were neither adopted nor produced during the war. Meanwhile, the rocket artillery that did exist during the Great Patriotic War, such as the BM-13, was not able to serve as an adequate substitute for tube artillery and had limited combat value.

Finally, despite the Red Army having accumulated a substantial amount of combat experience, it was not fully studied and implemented after the war. As a result, the armed forces of the Soviet Union, including tank and mechanized forces, continued their development in the general direction set during the Great Patriotic War, resulting in an incoherent armored forces system featuring numerous capability gaps. Similar to the pre-war USSR, the evolution of the armed forces was heavily influenced by the interference of the political leadership of the country and false assumptions about the nature of future conflicts. Of course, these core processes involved self-propelled artillery. In the post-war years, the existing fleet underwent a series of modernizations. However, most of the development programs initiated after the war were either canceled or postponed as the Soviet leadership shifted its focus to missile and rocket artillery.

Indirect-fire tube artillery capable of accompanying mechanized units and formations was not mass-produced in the 1940s and 1950s. It was only in the mid-1960s that the Soviet military realized that rocket artillery could not effectively accompany or support mechanized formations in modern warfare or efficiently accomplish the full spectrum of combat tasks envisaged by the ground forces. Remarkably, until the 1960s, self-propelled artillery was considered a variant of tanks, essential for accompanying tank units and primarily deployed in an anti-tank role.[239]

Only between the late 1960s and the beginning of the 1970s, the Soviet armored forces and self-propelled artillery achieved the level of the 1930s, following a comprehensive revision of the role of self-propelled artillery and the introduction of new equipment into service. Eventually, the perfect combination of massed artillery fires enabling armored maneuver was born and became the hallmark of the modern Russian Army.

The Photographs

Interwar

SU-18 self-propelled gun

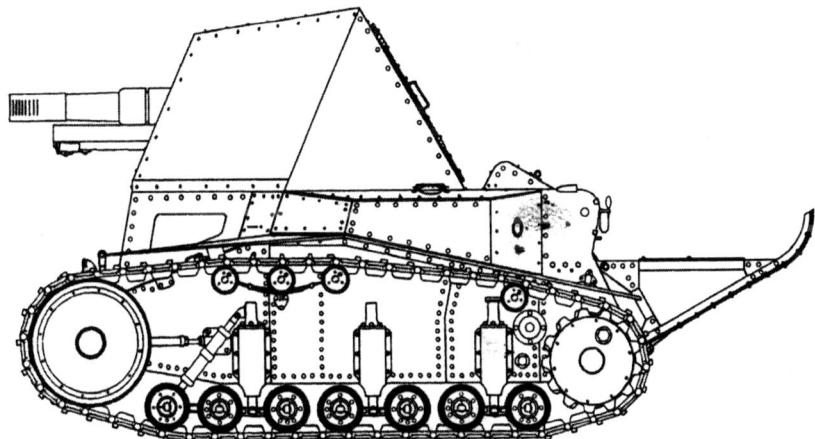

The SU-18 was one of the earliest attempts to develop a self-propelled gun, and the very first endeavor to build a family of armored vehicles on a common tracked platform. In December 1927 the ANII scientific-research design bureau started developing four different vehicles on an MS-1 tank chassis. The family of vehicles included two self-propelled artillery systems armed with 45mm and 76.2mm guns, a self-propelled air-defense vehicle armed with a twin 37mm gun mount, and another SPAAG armed with 7.62mm machine guns. Pictured: a drawing of the left side of the SU-18 SPG. (*Source: Author's collection*)

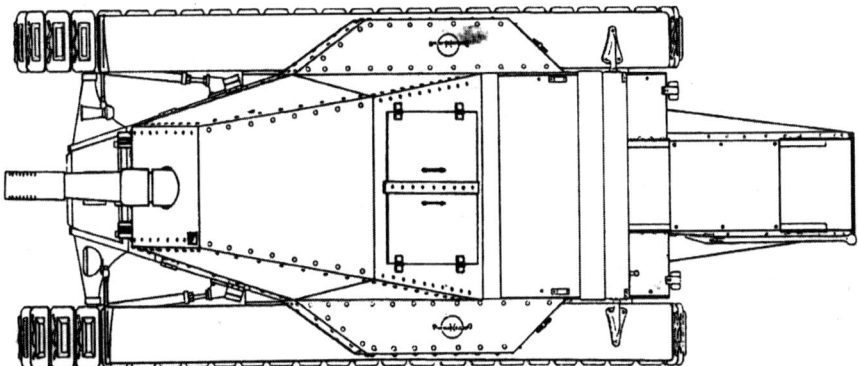

A drawing of the SU-18 SPG, top view. The vehicle retained the same arrangement as the MS-1 tank, with the driving compartment at the front, fighting compartment in the middle and engine at the rear of the vehicle. Due to the modest size of the hull, the SPG was able to carry only 4–6 artillery rounds inside. To address this issue, the ANII design bureau developed another vehicle on the same T-18 chassis — an armored ammunition transporter. (*Source: Author's collection*)

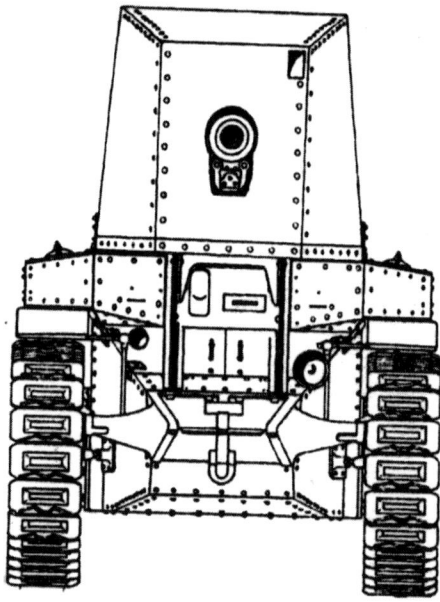

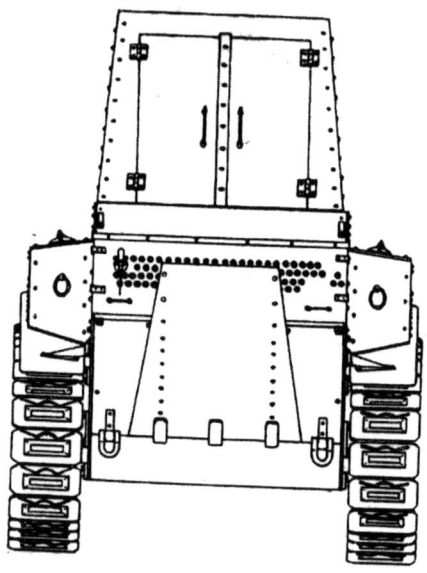

A front view of the experimental SU-18 self-propelled gun. Note the pyramid-shaped armored superstructure. The high center of mass was one of the obvious reasons for abandoning the SU-18 design. (*Source: Author's collection*)

A drawing of the SU-18 self-propelled gun, rear view. Eventually, only the SU-18 project was completed. The research and development works revealed that the 76.2mm M1927 regimental gun could not be fitted on a T-18 tank chassis without a major redesign of the base vehicle and it was decided to continue experiments on the bigger T-19 tank chassis. (*Source: Author's collection*)

SU-2 self-propelled gun

The SU-2 was another attempt to build a self-propelled gun, using proven technology. In fact, the SU-2 reproduces the German experimental self-propelled gun on a Hanomag WD 50 commercial tractor chassis also known as 7.7cm W.D. Schlepper 50PS. While the German SPG was developed in 1927, the Russian counterpart appeared in 1931. The SU-2 was built on a "Kommunar" tractor chassis, essentially the same Hanomag WD 50, and was armed with a 76.2mm M1902 field gun. The SU-2 had a crew of five, and the gun shield and hull were armored with 10mm thick plates. The vehicle carried ammunition on a special trailer and was able to reach the maximum speed of 12km/h. The prototype successfully passed the initial trials, but the further development was cancelled because the tractor chassis has proven itself too weak to endure the powerful 76.2mm gun. Note the Soviet slogan on the side: "We don't want war. But we are ready to fight back." (*Source: RGAE*)

There was another slogan on the other side of the vehicle: "Long live the world-wide October." (*Source: RGAE*)

SU-1 self-propelled gun

The SU-1 was an experimental self-propelled gun developed and designed on short notice in 1931. The vehicle was built on the T-26 light tank chassis and armed with a 76mm M1927 regimental gun or a 76mm M1902 divisional gun, as an option. In essence it was a crude testbed, hastily designed and assembled. While the vehicle demonstrated unsatisfactory results during trials in the fall of 1931, it was a good starting point for the further development of the self-propelled artillery. (*Source: RGVA*)

AT-1 artillery tank

The name AT-1 stands for artillery tank or "artilleriiskii tank" in Russian. This vehicle is often wrongfully considered as an assault gun, but in fact it was a double-role self-propelled gun capable to provide infantry with direct and indirect fire support. The ATs belonged to the class of the so-called "special purpose" or NPP tanks. In translation from Russian NPP means "neposredstvennoy podderzhki pekhoti" or direct infantry support. According to the Deep Battle doctrine the ATs were supposed to advance with the 3rd wave, providing infantry units with direct fire support, helping to mop-up enemy positions and destroying remaining centers of resistance. Pictured: an AT-1 prototype during trials in the winter of 1935. (*Source. Archive of the Museum of Artillery, Engineer and Signal Corps in St. Petersburg, Author's collection*)

A front view of the AT-1 during the winter trials in 1935. The AT-1 was developed by Design Bureau of the Experimental Specmashtrest's Plant in Leningrad in 1934. The SPG was armed with a powerful 76.2mm PS-3 gun and a 7.62mm DT machine gun. A second spare DT machine gun was carried inside. The project was led by Pavel Syachentov, a talented Russian engineer who also developed a variety of artillery systems and self-propelled guns including SU-5 and SU-14 families. (*Source. Archive of the Museum of Artillery, Engineer and Signal Corps in St. Petersburg, Author's collection*)

A rear view of the AT-1 with hatches and flap-doors closed. The AT-1 successfully passed factory and state trials and was approved for serial production in 1937. Initially the Armored Directorate wanted to acquire 100 ATs, however industry failed to deliver the order, and despite all efforts the AT-1 project was cancelled in August 1938. The only battle-ready prototype saw combat in 1941. (*Source. Archive of the Museum of Artillery, Engineer and Signal Corps in St. Petersburg, Author's collection*)

A view of the interior of the AT-1 (version 1936) through the opened rear flap-door. Left to right: 1—Commander's vision slit covered with the armored shutter; 2—opened hatch for the panoramic sight and panoramic sight; 3—gun telescopic sight (TOP); 4—PS-3 gun breech; 5—loader's periscopic sight; 7—a ball mount for the DT machine gun. (*Source: Archive of the Museum of Artillery, Engineer and Signal Corps in St. Petersburg, Author's collection*)

An AT-1 prototype with all hatches opened. While acting "buttoned up" in the direct fire support role, the commander was supposed to use the TOP telescopic sight, while the loader used the PTK periscopic sight. When firing indirectly, with flap-doors and hatches opened, the commander used the panoramic sight. Note the small hatch for the panoramic sight above the headlight. The PS-3 gun is at maximum elevation. (*Source. Archive of the Museum of Artillery, Engineer and Signal Corps in St. Petersburg, Author's collection*)

SU-5 "small triplex"

SU-5-1

A SU-5 or the "small triplex" ("Malii tripleks" in Russian) was an ambitious project of designing and building a family of self-propelled guns on a common T-26 tank chassis. The family of vehicles included three types of SPGs armed with different artillery systems: a 76.2mm L/30 gun, a 122mm L/12.8 howitzer and a 152.4mm L9.3 divisional mortar. The SU-5 family was designed in 1934 at Specmashtrest's design bureau. Pictured: a SU-5-1 self-propelled gun armed with a 76.2mm L/30 M1902/30 gun. (*Source: RGAE*)

A SU-5-1 light self-propelled gun seen from the left side. The gun is at maximum elevation of 60 degrees. All SU-5 family vehicles retained the engine, transmission and suspension of the T-26 light tank. In total six SU-5-1 were built before the Great Patriotic War. (*Source: RGAE*)

SU-5-2

A SU-5-2, a light SPG armed with a 122mm L12.8 M1910/30 howitzer. The SPG carried only 4 artillery rounds and 6 charges onboard and, therefore, needed an accompanying artillery transporter. However, the SU-5-2 needed only 1.38 to 1.54 minutes for transition from travelling to combat position and was able to fire at an average rate of 4 rounds per minute and 5–6 rounds per minute with a trained crew. (*Source: RGAE*)

A SU-5-2 with the howitzer at maximum elevation of 60 degrees. Remarkably enough, the SU-5 family guns were capable of opening fire almost immediately after the vehicle stopped, without the need to lower trails. (*Source: RGAE*)

The SU-5 family vehicles had a crew of five: a driver and a gun crew consisting of commander, gun-layer and two loaders. Pictured: a SU-5-2 in travelling position during trials. Note the opened driver's hatch and howitzer covered with tarpaulin. The gun crew is on their seats behind the gun shield. (*Source: Archive of the Museum of Artillery, Engineer and Signal Corps in St. Petersburg*)

A SU-5-2 prepares to fire during the live-fire tests. While the trials revealed several minor deficiencies, the performance of the gun was considered generally good and it was recommended for the army trials. The engineers noted good maneuverability and speed of the vehicle, as well as its accuracy and stability. The vehicle was adopted into service in 1937 and in total six SU-5-2 were built. The SU-5-2 saw combat in 1941. (*Source: Archive of the Museum of Artillery, Engineer and Signal Corps in St. Petersburg*)

The list of shortcomings mostly included ergonomics issues and a few mechanical flaws. Note that the gun is traversed to the extreme right position, and the loader standing to the left of the gun breech experiencing difficulties while loading the gun. (*Source: Archive of the Museum of Artillery, Engineer and Signal Corps in St. Petersburg*)

SU-5-3

The SU-5-3 was the variant armed with a 152.4mm M1931 divisional mortar. The mortar had a barrel length of 9.3 calibers and a maximum elevation of 72 degrees. The SPG was able to fire 4 rounds per minute at a maximum distance of 5285m. The crew was protected with a 15mm thick armored shield, however the commission recommended to design a superstructure with armored sides in order to protect the crew from flanking fire. A total of three SU-5-3 SPGs were produced. (*Source: RGAE*)

A SU-5-3 with the mortar at maximum elevation and lowered trails. (*Source: RGAE*)

SU-14

SU-14

The SU-14 was a family of experimental heavy self-propelled guns developed between 1933 and 1940. The design of the first variant incorporated some components from the T-28 medium tank and was armed with a 203mm M1931 B-4 howitzer. Pictured: the SU-14 prototype at the NIAP proving grounds during live-fire trials in 1935. The gun is at maximum elevation of 60 degrees. Note cranes and winches that were used to lift shells and charges from the ground. (*Source: Archive of the Museum of Artillery, Engineer and Signal Corps in St. Petersburg, Author's collection*)

The same vehicle with the gun in travel position. The trials revealed numerous ergonomic-related issues, which caused safety hazards for the crew, increased fatigue and reduced rate of fire. Another problematic issue was mobility. The maximum aimed rate of fire was only one shot per 5.7 minutes, three times slower than the rate of fire of the same 203mm howitzer in a towed variant. However, the commission admitted that during the live-fire trials crew was ordered to leave the deck and hide in the shelter nearby for safety reasons. (*Source: Archive of the Museum of Artillery, Engineer and Signal Corps in St. Petersburg, Author's collection*)

SU-14-1

Soviet engineers attempted to fix problematic issues in the design of the prototype. Whenever possible, the components of the T-28 tank were removed and replaced with more reliable and durable elements of the T-35 heavy tank. However, it was decided to build another prototype on the T-35A heavy tank chassis. Despite all the deficiencies, the SU-14-1 was accepted into service in June 1935. (*Source: RGAE*)

In 1934 the SU-14 became an artillery duplex able to carry both 203mm B-4 howitzer or 152.4mm B-10 special power gun. Later, in 1937 the SU-14-1 received another, more powerful gun—the B-30. Pictured is the later variant of the SU-14-1 SPG, armed with a 152mm B-30 gun. (*Source: RGAE*)

The SU-14-1 during live-fire trials in April–May 1936. This is the first prototype with improved design, but still armed with a 203mm B-4 howitzer. Note the tracks sunk in the mud. With a combat-ready weight of 47 tons, the vehicle had a ground pressure of 0.71–0.68 kg/cm². (*Source: Archive of the Museum of Artillery, Engineer and Signal Corps in St. Petersburg, Author's collection*)

SU-14-2 and SU-14-Br-2

SU-14-2 and SU-14-Br-2 SPGs were not designed from scratch, but converted from the SU-14 and SU-14-1 prototypes at the Factory #185. Both SPGs received new armor, fully enclosed superstructures and new armament, while engine, suspension, transmission and electrical system remained without any changes. Both self-propelled guns were delivered in 1940. (*Source: TsAMO*)

The SU-14-Br-2 was armed with a powerful 152.4mm Br-2 gun and received a fully enclosed casemate. The armor thickness for the front was 50mm, while the superstructure had 20mm thick armor. The SPG was intended to destroy Finnish fortifications and bunkers with direct fire, effectively to perform a heavy assault gun's task. (*Source: Author's collection, Kubinka Tank Museum*)

The vehicle had a weight of 65 tons and was able to travel with a maximum speed of 22km/h. The Winter War ended before the SU-14-Br-2 SPGs were ready for combat, however these vehicles saw limited military service during the Battle of Moscow in 1941. While the project was not entirely successful, it was a significant theoretical achievement. The SU-14 prototypes proved the possibility of building heavy-caliber self-propelled artillery using components of the serially produced tanks. (*Source: Author's collection, Kubinka Tank Museum*)

T-27

The T-27 light self-propelled gun was developed in 1931 at the Bolshevik factory on the T-27 tankette chassis. The vehicle was armed with a 73mm Hotchkiss gun and had minor improvements of the running gear. The machine-gun mount was eliminated, which allowed better elevation angles for the gun. In other aspects the vehicle was almost identical to the T-27M, another prototype designed by the Bolshevik factory. (*Source: RGAE*)

T-27M

The T-27M (also dubbed as T-27S) was another experimental project of the light self-propelled gun initiated at Bolshevik factory in 1931. Compared to the T-27 light self-propelled gun and serially produced T-27 tankette, this vehicle was armed with a 37mm Hotchkiss gun and a 7.62mm DT machine gun. There were also some minor differences in running gear and armored hull. (*Source: RGAE*)

The prototype was trialed in 1932, but never recommended for serial production due to the significant shortcomings in the design. For example, it was impossible to use a gun sight to aim the machine gun, while the armoring of the engine's radiator was designed poorly and caused overheating. Interestingly, that since the vehicle had a very limited inner space, it needed a specially designed tracked trailer to carry ammunition. (*Source: RGAE*)

SU-37 self-propelled anti-tank gun

The SU-37 was an experimental anti-tank SPG built on a T-37A amphibious light tank's chassis. In order to accommodate a 45mm anti-tank gun, the small rotating turret was removed and the gun was mounted in a fully enclosed fighting compartment. The SPG had a crew of two, the driver and the commander, who also acted as a loader, gunner and machine gunner. The SU-37 failed initial trials and the project was cancelled. (*Source: RGAE*)

SU-76 (T-27-based)

The SU-76 was an experimental 76mm self-propelled regimental gun on the T-27 tankette chassis. This vehicle was designed in 1932–34 by the "Krasny Putilovets" factory design bureau. The SU-76 was intended for mechanized infantry units as a mobile direct fire support vehicle. Only three prototypes were built in 1935. Pictured here is a combat-ready SU-76 with its trails lowered and main gun at maximum elevation. (*Source: Archive of the Museum of Artillery, Engineer and Signal Corps in St. Petersburg*)

As in the case with the SU-18 and SU-5, the SU-76 was an artillery complex consisting of two vehicles—a self-propelled gun and an armored ammunition transporter. Both were designed on the T-27 tankette chassis. (*Source: Archive of the Museum of Artillery, Engineer and Signal Corps in St. Petersburg*)

This photo shows the same SU-76 with its gun and trails in travel position. (*Source: Archive of the Museum of Artillery, Engineer and Signal Corps in St. Petersburg*)

SU-6 self-propelled AA gun

The SU-6 was designed in 1934 by engineer L. Troyanow, the project being supervised by P.N. Syachentov. The prototype was delivered in 1935 and tested in October–December 1936. In total, four SU-6 SPAAGs were delivered in January 1937 by Factory #185. (*Source: Archive of the Museum of Artillery, Engineer and Signal Corps in St. Petersburg*)

The SU-6 was designed as a vehicle able to accompany mechanized units and able to engage both airborne and ground targets. The SU-76 was armed with a 76mm 3K gun and two DT machine guns for self-defense. Pictured: the SU-6 during trials with its gun and armored shields in travel position. (*Source: Archive of the Museum of Artillery, Engineer and Signal Corps in St. Petersburg*)

This photo shows the SU-6 in ready-to-fire position with armored shields lowered and gun at an elevation of 45 degrees. The gun had a maximum elevation of 85 degrees and full 360-degree traverse. The SU-6 was built on the modified chassis of the T-26 tank, while retained the same engine, transmission, gearbox and other components. (*Source: Archive of the Museum of Artillery, Engineer and Signal Corps in St. Petersburg*)

SU-12

An iconic photo of the SU-12 wheeled self-propelled gun during the October Revolution Parade in Moscow, on November 7, 1936. There were two variants of SU-12 SPG. The initial variant was designated SU-12 and built on the US-made Moreland 6x4 truck chassis, while the late variant used domestically produced GAZ-AAA truck chassis and was named SU-1-12. The SU-12 was armed with a 76.2mm M1927 regimental gun and seen combat service at Khalkhin Gol and reportedly during the Winter War.

29K truck-borne AA gun system

29K or Motorized Anti-Air Gun System "29K" was a truck-borne anti-aircraft gun system based on a YaG-10 heavy truck chassis. The project was initiated in early 1930s, and the initial production batch of 20 vehicles was delivered in late 1936. A total of 40 vehicles (another source mentions 61) were produced between 1935 and 1937. Allegedly, some of the 29K SPAAGs participated in Great Patriotic War battles in 1941 in anti-aircraft and anti-tank roles. This photo shows a 29K in travel position with the gun covered by a tarpaulin (*Source: Archive of the Museum of Artillery, Engineer and Signal Corps in St. Petersburg*)

The 29K SPAAGs were designed as a corps-level air-defense asset for the Red Army's mechanized units. It was planned to replace all the towed AA-guns with "motorized" SPAAGS able to accompany mechanized troops and provide air defense against enemy aircraft. Pictured: a combat-ready 29K SPAAG with its 76.2mm 3K gun on maximum elevation. Note the crew lined up on the left. (*Source: Archive of the Museum of Artillery, Engineer and Signal Corps in St. Petersburg*)

1941–1943

ZiS-30

The ZiS-30 was one of the many makeshift SPGs, designed during the initial period of the Great Patriotic War. It was constructed on the base of a "Komsomolets" light artillery tractor and armed with a 57mm ZiS-2 gun. Initially, it was planned to build as many as 3000 ZiS-30 SPGs in just five months, but later this overly ambitious plan was cancelled. The production of "Komsomolets" tractors was discontinued in August 1941, and the supply of ZiS-2 guns was still insufficient. As a result, the Soviet industry struggled to produce only 100 self-propelled guns of this type. (*Source: Author's collection, UMMC Museum complex in Verkhnyaya Pyshma*)

ZiS-30 SPGs were intended for providing tank brigades with mobile anti-tank defense and supporting them in offensive actions. However, the combat employment of these SPGs produced very mixed results. On the one hand, ZiS-30 offered good maneuverability, excellent firepower and was easy to conceal due to its small silhouette. On the other, it had an underpowered engine, that limited vehicle's mobility on difficult terrain, in the snow or mud, and weak armor. Additionally, its engine and running gear were prone to break down during long marches, which was exacerbated with general lack of spare parts. Shown here is a replica of ZiS-30 painted in three-color camo pattern, typical for the vehicles produced in 1941. (*Source: Author's collection, UMMC Museum complex in Verkhnyaya Pyshma*)

Pictured here is a ZiS-30 self-propelled gun in position. The photo is likely staged, as the trails are not lowered. (*Source: TsAMO*)

29K truck-borne AA gun system

29K or Motorized Anti-Air Gun System "29K" was a truck-borne anti-aircraft gun system based on a YaG-10 heavy truck chassis. The project was initiated in early 1930s, and the initial production batch of 20 vehicles was delivered in late 1936. A total of 40 vehicles (another source mentions 61) were produced between 1935 and 1937. Allegedly, some of the 29K SPAAGs participated in Great Patriotic War battles in 1941 in anti-aircraft and anti-tank roles. This photo shows a 29K in travel position with the gun covered by a tarpaulin (*Source: Archive of the Museum of Artillery, Engineer and Signal Corps in St. Petersburg*)

The 29K SPAAGs were designed as a corps-level air-defense asset for the Red Army's mechanized units. It was planned to replace all the towed AA-guns with "motorized" SPAAGS able to accompany mechanized troops and provide air defense against enemy aircraft. Pictured: a combat-ready 29K SPAAG with its 76.2mm 3K gun on maximum elevation. Note the crew lined up on the left. (*Source: Archive of the Museum of Artillery, Engineer and Signal Corps in St. Petersburg*)

1941–1943

ZiS-30

The ZiS-30 was one of the many makeshift SPGs, designed during the initial period of the Great Patriotic War. It was constructed on the base of a "Komsomolets" light artillery tractor and armed with a 57mm ZiS-2 gun. Initially, it was planned to build as many as 3000 ZiS-30 SPGs in just five months, but later this overly ambitious plan was cancelled. The production of "Komsomolets" tractors was discontinued in August 1941, and the supply of ZiS-2 guns was still insufficient. As a result, the Soviet industry struggled to produce only 100 self-propelled guns of this type. (*Source: Author's collection, UMMC Museum complex in Verkhnyaya Pyshma*)

ZiS-30 SPGs were intended for providing tank brigades with mobile anti-tank defense and supporting them in offensive actions. However, the combat employment of these SPGs produced very mixed results. On the one hand, ZiS-30 offered good maneuverability, excellent firepower and was easy to conceal due to its small silhouette. On the other, it had an underpowered engine, that limited vehicle's mobility on difficult terrain, in the snow or mud, and weak armor. Additionally, its engine and running gear were prone to break down during long marches, which was exacerbated with general lack of spare parts. Shown here is a replica of ZiS-30 painted in three-color camo pattern, typical for the vehicles produced in 1941. (*Source: Author's collection, UMMC Museum complex in Verkhnyaya Pyshma*)

Pictured here is a ZiS-30 self-propelled gun in position. The photo is likely staged, as the trails are not lowered. (*Source: TsAMO*)

ZiS-41

The experimental ZiS-41 was another hasty attempt to build a self-propelled anti-tank gun. The vehicle was based on a ZiS-22M half-track truck and armed with a 57mm ZiS-2 gun and a DT machine gun. This photo shows a ZiS-41 at the proving grounds in April 1942. (*Source: TsAMO*)

The ZiS-41 was designed as a self-propelled anti-tank gun. It had a fully armored cab and a gun mount protected from the front and sides against small arms fire and artillery shell splinters. (*Source: TsAMO*)

This photograph shows the ZiS-41 overcoming a 15-degree front slope covered with deep snow. Generally, the vehicle demonstrated good off-road performance and ability to move through deep snow with average speed of 5–6km/h, off-road or through muddy terrain with average speed of 6–8km/h, while the average speed on a slope was 25km/h. (*Source: TsAMO*)

The ZiS-41 during live-fire trials: the gun is traversed to an angle of 90 degrees. Note the loader standing behind the shield in a dangerously insecure position. (*Source: TsAMO*)

The same live-fire trials, now the gun is traversed to 0 degrees. The ZiS-41 had ammunition stored on board, but the loader had to come out of the armored shield to take a new round. (*Source: TsAMO*)

Another version of the ZiS-41 tested in 1941 was an armored artillery tractor. Pictured here is a ZiS-41 towing a 122mm M-30 howitzer and limber. The wooden crates on the open platform imitate the weight of 80 artillery rounds with a total weight of 3 tons. (*Source: TsAMO*)

A ZiS-41 artillery tractor stands in the middle of a muddy road after the rain. A group of eight soldiers, imitating the gun crew, sits on the ammo crates. (*Source: TsAMO*)

A rear view of the ZiS-41 anti-tank variant with its gun mount dismantled. The photo offers a good view of the pyramid-shaped socket mount and platform. (*Source: TsAMO*)

The ZiS-41 had a crew of 5. The commander and driver sat in the armored cabin, the commander also operated the DT machine-gun. The loader and gun layer sat behind the gun shield the back. Finally, the observer sat behind the cabin on the open platform. Trials revealed numerous ergonomic issues. This photo shows vividly the cramped interior of the cabin and uncomfortable posture of the driver. Despite relatively good performance shown during trials, the project was cancelled for the same reasons as the ZiS-30: low production rate of ZiS-2 guns and unavailability of base chassis. The only prototype of the anti-tank SPG was converted to an armored artillery tractor. (*Source: TsAMO*)

ZiS-43

Another pressing issue for Soviet industry was developing mobile means of anti-aircraft defense for the tank formations. The ZiS-43 SPAAG was developed to fill this capability gap. Pictured here is a ZiS-43 prototype during trials in December 1942. The gun mount is covered with tarpaulin. (*Source: TsAMO*)

The ZiS-43 self-propelled anti-aircraft gun was based on an experimental ZiS-42 half-track chassis. The SPAAG was armed with a 37mm 61-K cannon and had no secondary armament. (*Source: TsAMO*)

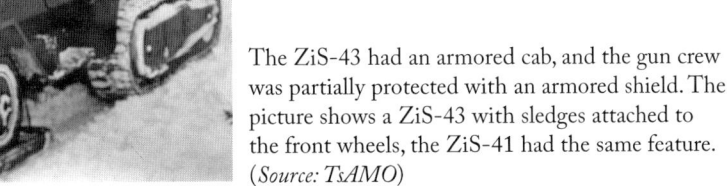

The ZiS-43 had an armored cab, and the gun crew was partially protected with an armored shield. The picture shows a ZiS-43 with sledges attached to the front wheels, the ZiS-41 had the same feature. (*Source: TsAMO*)

The ZiS-43 had a crew of 8 and had a combat-ready weight of 8750kg. Pictured here is a ZiS-43 during trials with its gun at maximum elevation of 85 degrees. The 61-K gun had an average rate of fire of 73 rpm (rounds per minute). (*Source: TsAMO*)

Probably the most significant problem with both the ZiS-41 and ZiS-43 half-tracks was their underpowered engine and its inadequate air cooling, which caused frequent engine overheating, increased fuel consumption and limited mobility. The ZiS-43, for example, was able to reach the maximum speed of only 35km/h on a paved road. Additionally, the trials revealed that temperature inside the cabin reached +40–50 °C. (*Source: TsAMO*)

Despite the deficiencies revealed during trials, the commission gave the green light to the ZiS-43. The SPAAG was recommended for adoption into service with the Red Army, however, the GABTU (Main Automotive-Armored Directorate) decided to cancel the project. The reasons were roughly the same as in case with the other experimental vehicles: scarce resources and insufficient industrial capacity. This photo gives a clear view on the vehicle's rear, note the sledges attached to the back side. (*Source: TsAMO*)

Armored ZiS-5

Among the lesser-known combat vehicles of the Red Army, there is the ZiS-5 "Model PB" wheeled SPG. The acronym "PB" stands for "pushechnyi bronirovannyi" or armored, with a gun. In fact, it was a commercial truck armored and armed with a 45mm gun. The improvised combat vehicles of this type are known today as "guntrucks". The armament varied greatly from production batch to production batch. There were variants armed with 45mm guns, 20mm ShVAK autocannon, and various machine guns. (*Source: TsAMO*)

The reason behind developing this vehicle was to offer more mobility to towed artillery systems used for direct infantry support. While the civilian chassis allowed to quickly convert trucks to combat vehicles, the efficiency of the vehicle was limited. The chassis and engine of a commercial truck were not supposed to endure additional weight of armor and armament. The increased load caused frequent mechanical breakdowns and limited mobility. (*Source: Author's collection, UMMC Museum complex in Verkhnyaya Pyshma*)

The ZiS-5 guntruck was designed in July 1941 at Izhorsky plant near Leningrad. The self-propelled gun featured a fully armored cabin, offering protection against small arms fire and shell splinters. The gun mount was partially protected from the front and sides. (*Source: Author's collection, UMMC Museum complex in Verkhnyaya Pyshma*)

Armored GAZ-AA

Another model designed and produced at Izhorsky plant was a GAZ-AA "Model ZP". "ZP" stands for "zenitno-pulemetnyi" or anti-aircraft machine gun. The truck-based improvised armored vehicle was armed with a twin machine-gun mount, and had a fully protected engine and armored cabin. The wooden cargo bay was partially protected from the sides. It is unclear how many guntrucks and their variants were manufactured. However, the surviving documents suggest that the Izorsky plant produced 300 "PB" and "ZP" guntrucks in July 1941 alone.

Object 212

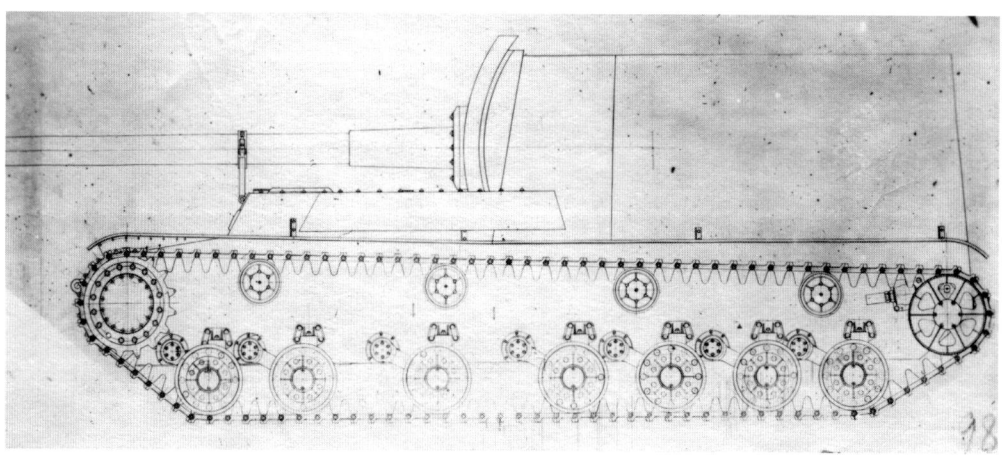

A left-side view drawing of the "Object 212" heavy self-propelled howitzer. The project was initiated in August 1941, continuing with numerous ups and downs till early 1942, and eventually was cancelled. The development was led by leading engineer Golzgurg and supervised by Zhores Kotin. The experimental 65-ton self-propelled howitzer inherited elements and solutions used on SMK heavy tank and KV-family tanks. It was powered with a V2 engine, armed with a 152.4mm BR-2 howitzer and had a crew of 7.

KV-7

The KV-7 was one of the first attempts to develop a tank-support combat vehicle. Officially it was designated as a heavy tank, but in fact it was a typical self-propelled gun with a casemate fixed superstructure which could be considered the first Soviet heavy self-propelled gun. The most prominent feature of this experimental armored vehicle was its armament. The first prototype was armed with two 45mm and one F-34 76mm gun, while the second with a twin 76mm gun mount. This photo shows a KV-7 prototype with a triple gun mount.

A clear view of the triple gun mount. The 76mm gun is mounted in the center, while 45mm guns are fitted to the left and right of it. The guns were able to fire in three combinations: salvo fire from three guns, salvo fire from any two guns or single fire from any gun.

The right side of the KV-7 prototype. The vehicle was built on the KV-1 heavy tank chassis and retained the same V-2K engine and transmission.

The experiment failed for various reasons. Firstly, the vehicle had a crew of 6 and required two loaders. Secondly, it was hard to maintain because the KV-7 required two different types of ammunition, gun sights and spare parts for the guns. And finally, all three guns were mounted in one cradle. This design did not allow to aim each gun separately. Moreover, due to the recoil the gunner had to adjust the horizontal aiming after each shot.

While the idea to arm combat vehicles with twin gun mounts was shelved until the end of the Great Patriotic War, it was not abandoned completely. Soviet industry revisited this design again, in the 1950s. Pictured: Front-right view of the KV-7 prototype.

SU-71

The SU-71 self-propelled assault gun was designed and manufactured at Gorkovsky automotive plant (GAZ) in 1942. In total two prototypes were produced, one of them was trialed from December 15 to 22, 1942 at the Gorokhovetsky proving grounds. Pictured is the SU-71 during trials, winter 1942. (*Source: TsAMO*)

The design of the SU-71 SPG included parts and engineering solutions used in the T-70 light tank. For example, it had a power plant consisting of two GAZ-202 liquid-cooled engines, developing 140 hp. The photo shows the SU-71 prototype with the gun at maximum elevation. (*Source: TsAMO*)

The photo shows the rear-right part of the SU-71 during the winter trials. The rectangular door on the back side of the superstructure was used for crew access as well for loading and unloading ammunition. Despite the visual impression that the SPG has a fully armored superstructure, it was, in fact, open-topped. The height of the fighting compartment was only 1370mm, which offered protection for the crew only when they were working crouched or on their knees. (*Source: TsAMO*)

The SU-71 was armed with the renowned 76mm ZiS-3 M1942 gun and had no secondary armament. It was a common issue for the Soviet self-propelled artillery, which remained mostly unresolved until 1944–45. Despite the SU-71 performing well during factory trials, its running gear failed completely during trials at Gorokhovetsky proving grounds. The other problems were typical for the Soviet armored vehicles of the time and included poor ergonomics and inadequate protection for the crew, to name a few. The commission concluded that SU-71 failed the trials, and the further development of the design was "impractical." (*Source: TsAMO*)

SU-76 (SU-12) M1942

The SU-12 self-propelled assault gun was developed in October–November 1942, at Factory #38's design bureau in Kirov. Pictured: a SU-12 self-propelled gun produced in March–June 1943. This vehicle has a fully enclosed fighting compartment with an armored roof, while the prototypes were open-topped. (*Source: TsAMO*)

The SU-12 was another design that was based on the T-70B light tank chassis. The SPG was armed with a 76mm ZiS-3 gun. (*Source: TsAMO*)

Despite numerous deficiencies, the SU-12 was accepted into service in December 1942 and remained in serial production until July 1943. In total, 560 (609 according to some sources) SU-12 self-propelled guns were manufactured. Serially-produced SU-12 SPGs suffered from various production and design defects. Probably the most serious of them were poor quality of assembly and faulty gearbox. Nevertheless, it was a long-awaited combat vehicle. The first units armed with SU-12 assault guns went to combat in February 1943. The new SPGs proved themselves well and received mostly positive assessment from the crews. (*Source: TsAMO*)

SU-76M (SU-15)

SU-15, the first prototype

The SU-15 was another T-70-based design, that initially appeared as a possible answer for the problems experienced by the SU-12 program. Eventually, the SU-15 evolved into the SU-76M, which became the most mass produced self-propelled gun of the Red Army. Pictured: the first prototype of the SU-15 SPG at the Gorokhovetsky proving grounds, June 1943. (*Source: TsAMO*)

One of the four SU-15 prototypes with the number 15-601 clearly seen on the side of the superstructure. The gun is at maximum elevation of 29 degrees. The SU-15 had demonstrated superb results during live-fire trials, but failed the mobility tests. While the Soviet engineers had been working hard at improving the designs, the running gear remained the weak spot of the Soviet self-propelled artillery till the end of the Great Patriotic War. (*Source: TsAMO*)

The same vehicle with rear doors opened. The prototypes had a fully-enclosed fighting compartment with a roof. In comparison with the SU-12, the engine compartment and interior of the fighting compartment were heavily redesigned. GABTU believed that increased wear and mobility problems were caused by the excessive weight, and in 1943 ordered to reduce armor thickness at the front and sides, as well as completely removing the armored roof and rear armored wall of the superstructure. The modified vehicle was designated SU-15M. (*Source: TsAMO*)

SU-15M, the second prototype—pre-series vehicle

The open-topped SU-15M was significantly lighter, weighing just 10.5 tons. The State Defense Committee (GKO) also approved to reduce the maximum speed to 30km/h. However, these changes came at the price of reduced protection and mobility, making the SPG an easy target even for mortar and machine gun fire. Pictured is the SU-15M with a serial number L57406 ("Л57406") during trials in August 1943. Note the designation "СУ-76М" written on the glacis plate, to the left of the driver's hatch.

Another SU-15M prototype with a serial number L57403 ("Л57403") seen from behind. The photo was taken during the comparative trials at the Gorokhovetsky proving grounds in August 1943. Both SU-15M and SU-16 prototypes completed the trials, but eventually the commission recommended the SU-15M for serial production as more advantageous. One of the unequivocally acknowledged benefits, in comparison with SU-16, was its spacious fighting compartment.

SU-76M (SU-15), *series model*

Throughout the war, SU-76M SPGs were produced at three factories: Factory #38 in Kirov, Factory #40 in Mytischi near Moscow and at GAZ in Gorky (Nizhny Novgorod today). Pictured: a serial SU-76M produced in 1944 at GAZ factory. The patterns of serially produced SU-76M SPGs varied from one factory to another.

The same vehicle, taken from the left side. The process of improvement was continuous and conducted at remarkably high pace. Despite that, the SU-76M suffered from numerous design and manufacturing defects till the end of production in October 1945.

This photo gives a good view on the general arrangement of the SPG and its open-topped fighting compartment. The SU-76M was extremely vulnerable to small arms fire from the sides and rear, as well as shell splinters from above. The inherent weaknesses were worsened by inadequate combat employment, which often led to high casualties among light self-propelled artillery units.

SU-15M, the second prototype—pre-series vehicle

The open-topped SU-15M was significantly lighter, weighing just 10.5 tons. The State Defense Committee (GKO) also approved to reduce the maximum speed to 30km/h. However, these changes came at the price of reduced protection and mobility, making the SPG an easy target even for mortar and machine gun fire. Pictured is the SU-15M with a serial number L57406 ("Л57406") during trials in August 1943. Note the designation "СУ-76М" written on the glacis plate, to the left of the driver's hatch.

Another SU-15M prototype with a serial number L57403 ("Л57403") seen from behind. The photo was taken during the comparative trials at the Gorokhovetsky proving grounds in August 1943. Both SU-15M and SU-16 prototypes completed the trials, but eventually the commission recommended the SU-15M for serial production as more advantageous. One of the unequivocally acknowledged benefits, in comparison with SU-16, was its spacious fighting compartment.

SU-76M (SU-15), *series model*

Throughout the war, SU-76M SPGs were produced at three factories: Factory #38 in Kirov, Factory #40 in Mytischi near Moscow and at GAZ in Gorky (Nizhny Novgorod today). Pictured: a serial SU-76M produced in 1944 at GAZ factory. The patterns of serially produced SU-76M SPGs varied from one factory to another.

The same vehicle, taken from the left side. The process of improvement was continuous and conducted at remarkably high pace. Despite that, the SU-76M suffered from numerous design and manufacturing defects till the end of production in October 1945.

This photo gives a good view on the general arrangement of the SPG and its open-topped fighting compartment. The SU-76M was extremely vulnerable to small arms fire from the sides and rear, as well as shell splinters from above. The inherent weaknesses were worsened by inadequate combat employment, which often led to high casualties among light self-propelled artillery units.

SU-76BM (NATI-TsAKB)

The experimental SU-76BM self-propelled gun was an initiative project to build a light SPG capable of supporting infantry, as well as fighting enemy armor and field fortifications.

The project was initiated in July 1943 and developed by NATI in cooperation with TsAKB. The prototype was delivered in October 1943 and tested at the Sofrino artillery range. While the SU-76BM passed factory trials, the project was cancelled due to numerous shortcomings in the design.

The SU-76BM was armed with a 76.2mm F-34 gun and had a maximum height of 1.68m. As noted in the official description, the SPG had "a height of a field artillery gun." The low profile of the vehicle contributed to better concealment and hence better survivability on the battlefield. On the flipside, the SU-76BM had a cramped fighting compartment making it difficult for the crew to operate.

The same prototype, seen from the rear. Note the commander's cupola, an unusual feature for the most Soviet light self-propelled guns designed in 1941–43. The hinged door has a firing port where a PPSh submachine gun could be placed. Like many other Soviet SPGs, the SU-76BM had no secondary armament.

An SU-76BM drives up a slope during trials. The vehicle was powered with two M-1 engines, had two gearboxes used on GAZ-AA family trucks, and other elements used on T-80 light tanks, such as running gear, brakes and side clutches.

SU-38

The SU-38 was one of many experimental projects for a light self-propelled gun developed at the GAZ plant in 1942–43. All designs, including the SU-38, SU-74 family, SU-71 and others, used serially-produced T-70 tank components. The SU-38 had a crew of 4, combat-ready weight of 10,600kg and armed with a 76.2mm ZiS-3 gun.

The same vehicle seen from the right side. While the early prototypes had a fully enclosed fighting compartment, the later version was open-topped. After several rounds of trials, SU-38 prototypes were heavily redesigned and re-designated as SU-16. The SU-16 became the main competitor to the SU-15M (later the SU-76M), but eventually lost the competition.

This photo shows the SU-38 prototype with a serial number "38602" painted on the side of the superstructure. The rear door is open wide.

SU-74B (SU-57B) M1943

The SU-74B was developed in June 1943 at the GAZ automotive plant design bureau under N.A. Astrov's supervision. This project pursued the same goal as SU-38 or SU-15—to develop a cheap combat vehicle based on serially produced components in order to saturate the Red Army with light self-propelled artillery on short notice.

The SU-74B prototype during trials in August–September 1943. The photo gives a good view on the prototype's suspension and the shape of the hull. The SU-74B was armed with a 57mm ZiS-4 anti-tank gun.

This photo shows the same vehicle from another angle. The details of the engine deck are clearly visible. The SU-74B was powered with a ZiS-16F carburetor engine able to produce 104 hp. In fact, the faulty engine and the inability of the plant to improve its performance were the main reasons behind the cancellation of this project.

SU-74D (SU-76D) M1943

The SU-74D was another experimental design that used the altered armored hull of the SU-74B, but was powered with an American GMC 4-7 diesel engine 1. Unlike the SU-74B, the SU-74D was armed with a 76.2mm F-34 gun and also had a redesigned transmission.

The same prototype seen from the right side. The SU-74B and SU-74D can be easily distinguished by their guns, gun mantlets and the shape of the armored hulls.

This photo was apparently taken at the NIBT proving grounds in August–September 1943 and gives a good view of the vehicle's roof. Despite the fact that the SU-74D successfully passed the initial trials, the design was rejected mainly due to its American-made engine, which, in commission's opinion, could hamper the mass production. Since Soviet industry failed to develop compact and reliable engines suitable for light combat armored vehicles, the project was cancelled.

SU-85A (SU-15A) M1943

After the Battle of Kursk in 1943 it became apparent to Soviet command that 76mm guns are not capable enough to fight with new German tanks. The obvious solution was to rearm the existing SPGs with more powerful guns, which was the main reason behind the SU-85A project, that was initiated in 1943 at the GAZ design bureau.

In essence the SU-85A was a basic SU-76M with a new 85mm D-5-S85 ("Д5-С85") gun and a few minor improvements.

The prototype was assembled between September and November 1943. While the combat weight increased to 12.3 tons, it did not affect vehicle's mobility or mechanical reliability. However, the trials revealed that the new gun had a powerful recoil. The comparative test revealed that an experimental SU-85A (based on SU-76M) had accuracy three times lower, compared to the serially produced SU-85 based on T-34 chassis.

A SU-85A prototype during tests at the NIBT proving grounds in December 1944. The gun is at maximum elevation, and the fighting compartment is covered by a tarpaulin. Compared to the SU-76M, the SU-85A was slightly longer (4.99m) and wider (2.755m), but had the same height of 2.1m.

SU-122

Prototypes

The prototype of the 122mm SPH designed by Factory #592. Note the running gear featuring all-metal wheels. Two prototypes were designed at UZTM and Factory #592. From December 5 to 19, 1942 they underwent comparative trials at the Gorokhovetsky proving grounds. (*Source: TsAMO*)

The same prototype seen from the front. The SPH has noticeable differences from the UZTM design, such as different design of the gun mount, driver's hatch and commander's cupola. Also the prototype has a machine-gun mount above the driver's hatch. (*Source: TsAMO*)

The prototype of the 122mm SPH designed by UZTM. The vehicle has different running gear featuring rubberized road wheels. The commander's cupola has a different shape and is offset to the rear of the superstructure. (*Source: TsAMO*)

A front view of the UZTM prototype. Note the size of the driver's hatch to the left of the gun. Crew used the rectangular hatch on the roof to access their seats, but this hatch was used by the driver for observation only. (*Source: TsAMO*)

U-35 (SU-122) the first vehicle of the pilot production batch

The photo shows one of the self-propelled guns of the pilot production batch in the UZTM factory yard, probably in the winter of 1942–43. These SPGs had a factory designation U-35. The final version of the U-35 was, in fact, a result of combined efforts of two design bureaus. While UZTM introduced a T-34-based prototype that completed live-fire and mobility trials in December 1942, the commission rejected it because of the poorly designed fighting compartment. Factory #592 introduced a prototype, that failed mobility tests, but had a much better arrangement of the fighting compartment. Since Factory #592 had experience with converting German combat vehicles to SPGs, it is safe to suggest that the internal arrangement was inspired by the StuG or Panzer III. (*Source: TsAMO*)

The same vehicle seen from another angle. The U-35 was built on a T-34-76 chassis and armed with an 122mm M-30 M1938 howitzer mounted in a fully enclosed fighting compartment. This medium SPG was intended for direct infantry support. In other words, the Soviet political and military leadership returned to the same pre-war ideas they rejected in 1937–40. (*Source: TsAMO*)

The left side of the U-35 SPG, showing details of the superstructure and suspension. The first prototype was delivered in November 1942, tested in December and accepted into service on December 5, 1942. Serial production started the same month. (*Source: TsAMO*)

The right side of the U-35 with its gun depressed at 3 degrees. UZTM delivered the pilot production batch of 25 SPGs by January 1943, and on January 3rd they were moved by train to Moscow. (*Source: TsAMO*)

Serially-produced SU-122

A SU-122 with serial number U-304272 ("У-304272") produced in May 1943. Serial vehicles had some differences compared to the pilot model, such as a small hatch on the lower front plate, allowing to access a track tension mechanism, and new design of the cupola for the panoramic sight. (*Source: TsAMO*)

A front view of the same SU-122. Note the commander's sight and details of the armored recuperator housing. (*Source: TsAMO*)

This photo was taken during SPH's warranty testing conducted from May till August 1943. While the serially produced SU-122 remained unreliable and retained numerous inherent design defects, it was highly praised by Soviet soldiers and commanders as an extremely useful and effective weapon. (*Source: TsAMO*)

The SU-122 had a crew of 5. Three crew members (driver, gunner and loader) sat to the left of the gun, while the commander and breechblock operator sat to the right. Since the 122mm M-30 originated from a towed howitzer, it retained its separate-loading ammunition and separate gun laying controls, with the traverse mechanism located on the left and elevation mechanism on the right. Therefore, the gunner and commander had to synchronize their actions; plus, the commander had an additional task in combat. As a result, accuracy and rate of fire were low and deemed "unsatisfactory." The average rate of fire was 5 rounds per minute from a stationary position, while crew needed 60–75 seconds to reload the gun. Rate of fire from a short stop was even lower—2–3 rounds per minute. (*Source: TsAMO*)

This photo gives a good view on a rear part of the engine compartment and superstructure. Note the handrail on the rear side of the superstructure. This feature would be eliminated in the late production batches. In addition to other defects, the SU-122 had no secondary armament. This, in combination with poor visibility, made the SPG "blind" and defenseless in close combat. (*Source: TsAMO*)

The production of the SU-122 was discontinued in August 1943 with a total of 640 vehicles produced. Pictured: a SU-122 SPG gets a new white paint job at one of the factories in Leningrad in 1944. (*Source: Author's collection*)

SU-122M

Despite the enormous list of defects, the SU-122 was pushed into serial production in January 1943. However, the engineers realized the need to improve the design and started working on the improved version almost immediately. Pictured: a SU-122M prototype at the UZTM plant in May 1943. (*Source: TsAMO*)

The prime area of concern was the main armament. In order to improve ergonomics and, hence, accuracy and rate of fire, the UZTM suggested to implement a new 122mm howitzer. The proposed system was modernized U-11 gun, later re-designated as D-11. Pictured: the same vehicle seen from the right side, gun at maximum elevation. (*Source: TsAMO*)

The new howitzer was more compact and weighed less than the original M-30. This allowed the combat compartment to be rearranged, eliminating hazards for the crew and generally improving ergonomics and habitability. The same vehicle, howitzer at maximum depression. (*Source: TsAMO*)

The same SPG seen from the front. Note a new less bulky gun armoring, allowing better field of vision to the driver, a bigger driver's hatch and a new telescopic sight. (*Source: TsAMO*)

Rear view of the SU-122M at the UZTM plant. The prototype was delivered only by late April 1943, completed factory trials in May, and was handed over to the state trials. (*Source: TsAMO*)

Top-front view of the SU-122M, the differences in the design of the casemate, the new gunner's hatch and armoring of the gun mount are clearly visible. (*Source: TsAMO*)

A side schematic view of the SU-122M. In comparison with the serial SU-122, the upgraded version was 50mm higher.

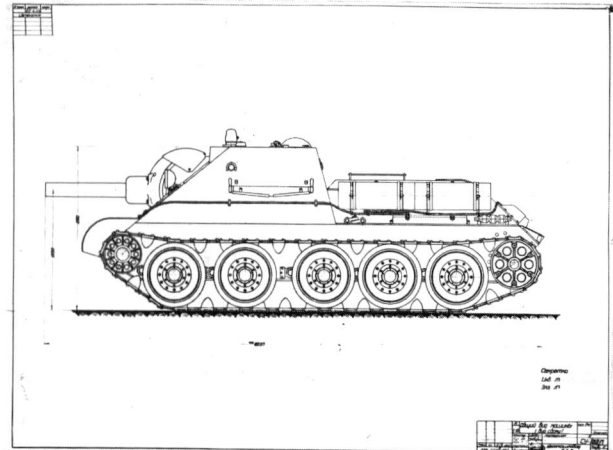

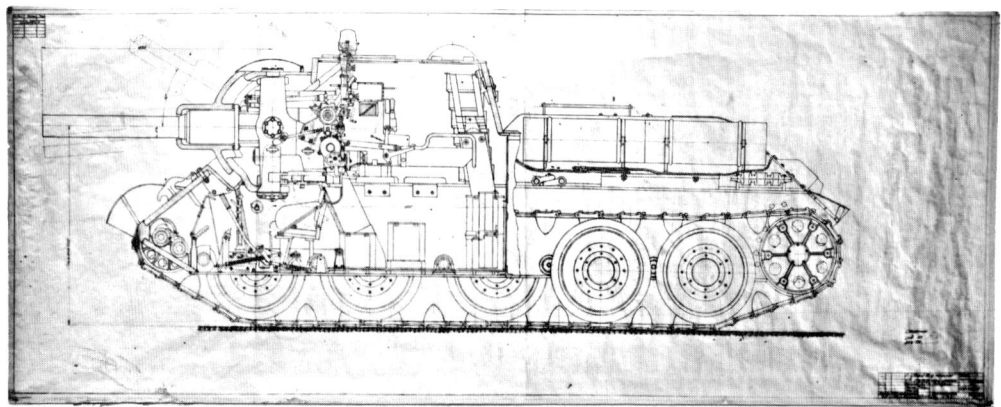

Cutaway of the SU-122M, showing the internal arrangement. The commander and breechblock operator received new seats, while the ammunition stowage was completely redesigned. Probably the most significant improvement was the new gun, requiring only the gunner to operate it. While the problems with ergonomics and situational awareness were not resolved completely, the SU-122M was accepted and recommended for the serial production in June 1943. However, the project was sidelined because the Soviet command wanted new SPGs able to confront new German tanks and self-propelled guns with thicker armor. (*Source: RGAE*)

SU-122-III

The SU-122-III project was conceived at Factory #9 (part of UZTM) in spring 1943 and was being developed in parallel with the SU-85 — the new top priority for the Soviet tank industry. Pictured: a front view of the SU-122-III prototype at the factory yard. Note that SU-122-III was a factory designation, while sometimes this vehicle was also designated as D-5-SU-122. (*Source: RGAE*)

The prototype was delivered in July 1943 and tested on 25 July along with three prototypes of the SU-85. Trials revealed a number of deficiencies, including troubles with the recoil mechanism. The latter issue resulted in deformation of some elements of the gun mount. Consequently, the commission removed the SU-122-III from trials. After the trials, the commission asked the factory to fix the defects, but it was preoccupied with highly demanded SU-85 tank destroyers, and the project was abandoned. Pictured: the same prototype with D-5-S-122 gun at maximum elevation. (*Source: RGAE*)

A right-side view of the SU-122-III. This version has a redesigned commander's cupola with a periscopic sight on it, larger driver's hatch and simplified design of the gun mantlet. (*Source: RGAE*)

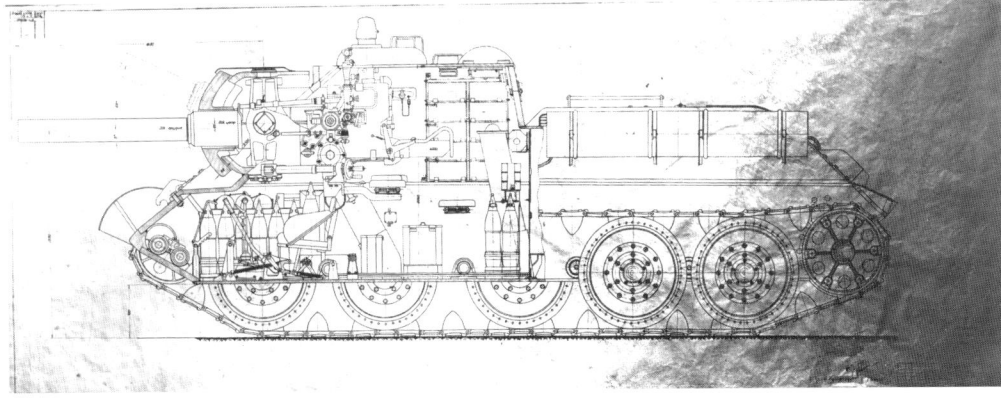

A cutaway of the SU-122-III SPG armed with a 122mm D-5-S-122 gun, June 1943. The D-5-S-122 was in fact a duplex allowing to change 85mm and 122mm barrels with minimal changes. This solution allowed to produce two kinds of SPGs (tank hunter and direct fire support) on the same chassis with negligible effort and changes in design. (*Source: RGAE*)

SG-122 (A)

The SG-122 was a medium self-propelled gun produced in small numbers throughout 1942–43. The SPH was built on captured German StuG III chassis, but was armed with a powerful M-30 howitzer and had a larger superstructure. Pictured: a replica of an SG-122 self-propelled gun at the UMMC Museum complex in Verkhnyaya Pyshma. (*Source: Author's collection, UMMC Museum complex in Verkhnyaya Pyshma*)

A total of 21 SG-122 SPGs were produced from October 1942 to January 1943. The SPHs saw limited service with the 1435th SAP (self-propelled artillery regiment). While the production was cancelled, the engineers received invaluable experience and knowledge. Some technical solutions were later implemented in other projects, such as the SU-122 and SU-152. (*Source: Author's collection, UMMC Museum complex in Verkhnyaya Pyshma*)

SU-85

Variant	Armament	Weight, tons	Status
SU-85-I	S-18-I with changes requested by the UZTM	29.66 (29.8)	Failed
SU-85-II	D-5S-85, factory #9's design	29.15	Accepted
SU-85-IV	S-18 by TsAKB	30	Failed

SU-85-I

To speed up the development of the new self-propelled anti-tank guns, all three prototypes (SU-85-I, SU-85-II and SU-85-IV) were built on the SU-122 chassis. The prototypes differed mainly by their armament and arrangement of the fighting compartment. Pictured: a front view of the SU-85-I. This variant was armed with an 85mm S-18-1 gun designed by TsAKB, with UZTM's changes. Note the small access hatch on the gun shield. (*Source: RGAE*)

The same vehicle seen from the left side. The most significant detail of the exterior, allowing to recognize this version, is the gun shield. (*Source: RGAE*)

The same SU-18-I, seen from the right side. This version weighed 29.66 tons (29.8 tons in some sources). (*Source: RGAE*)

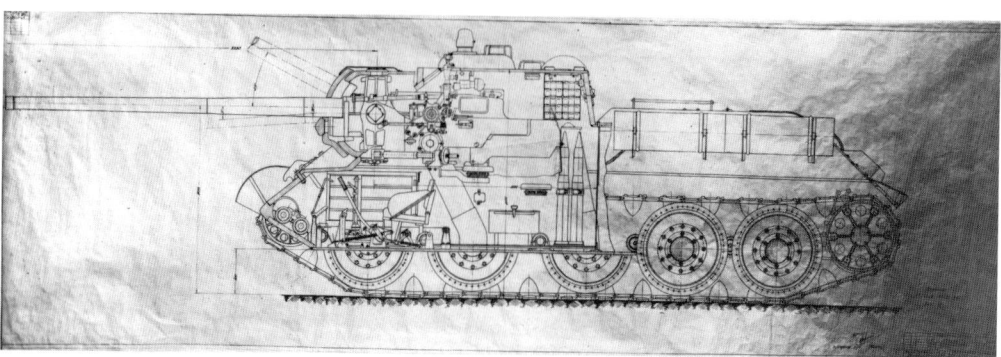

Cutaway of the SU-85-1, showing its internal arrangement. The fighting compartment of the SU-85 family vehicles was more spacious, thanks to the compact gun system and more organized internal space. (*Source: RGAE*)

The SU-85-I prototype during state trials at the Gorokhovetsky proving grounds, probably in July or August 1943. The details of the gun shield and commander's cupola are clearly visible. (*Source: RGAE*)

The same prototype seen from the left side. Note the "1-A" painted on the side of the superstructure. (*Source: RGAE*)

The SU-85-I prototype seen from the rear; note the unusual arrangement of the external fuel tanks attached to the sides. All three SU-85 prototypes had on-road and off-road mobility almost identical to that of T-34 medium tanks, with average speed of 22.6km/h on road and 18km/h off-road. (*Source: RGAE*)

The SU-85-I prototype at the Gorokhovetsky proving grounds, seen from the rear-right side. It appears from this angle that the vehicle carries five fuel tanks instead of the regular four. The SU-85 prototypes had 11 fuel tanks with a total fuel capacity of 820 liters, allowing a maximum range of 400km. (*Source: RGAE*)

SU-85-II

The SU-85-II was armed with a 85mm D-5S-85/D-5S-85A gun developed and designed at Factory #9. The SPG had a combat weight of 29.15 tons and was the lightest of the three prototypes that participated in comparative trials. (*Source: TsAMO*)

A left-side view of the SU-85-II in the factory yard. All three prototypes were powered with a V-2-34 engine developing 420 hp at 1700 rpm. (*Source: RGAE*)

Another shot of the SU-85-II in the factory yard with the driver's hatch closed. Note the details of the gun shield. The photo was taken in July 1943. (*Source: RGAE*)

The same vehicle with the hatches open. The driver's hatch was almost identical to the T-34's hatch. (*Source: RGAE*)

The SU-85-II at the Gorokhovetsky proving grounds in July–August 1943. Trials demonstrated the unambiguous superiority of the D-5S-85 system. It was the most lightweight, reliable and offered the highest rate of fire of 8 rpm. (*Source: RGAE*)

The same vehicle shown from the rear. Also the commission mentioned that the D-5S-85 had the shortest recoil and was the most compact of all systems tested. The latter allowed the best internal arrangement for the fighting compartment and, correspondingly, the best working conditions for the gunner and loader. (*Source: RGAE*)

The SU-85-II prototype seen from the rear right. SU-85 SPGs had no secondary armament, so two PPSh submachine guns were stored inside the fighting compartment. (*Source: RGAE*)

After the tests, the commission concluded that all four prototypes (three SU-85s and the SU-122-III) successfully completed mobility tests, but failed live-fire trials. However, the D-5S-85 demonstrated the best performance and reliability and was recommended for service with the Red Army. Pictured: the SU-85-II prototype seen from another angle. (*Source: RGAE*)

SU-85-IV

The SU-85-IV had the same arrangement of the fighting compartment as the SU-85-I, but was armed with a heavier 85mm S-18 gun designed by TsAKB. This photo shows the front view of the SU-85-IV at the factory yard. (*Source: RGAE*)

This variant is easily recognizable thanks to its bulky gun mantlet. Moreover, the S-18 gun by TsAKB was the heaviest system of the three introduced to the trials, weighing 1310kg without the gun shield. The SU-85-IV, therefore, was the heaviest prototype, weighing 30 tons. (*Source: RGAE*)

The right side view of the same SU-85-IV prototype with the gun depressed to 4 degrees. (*Source: RGAE*)

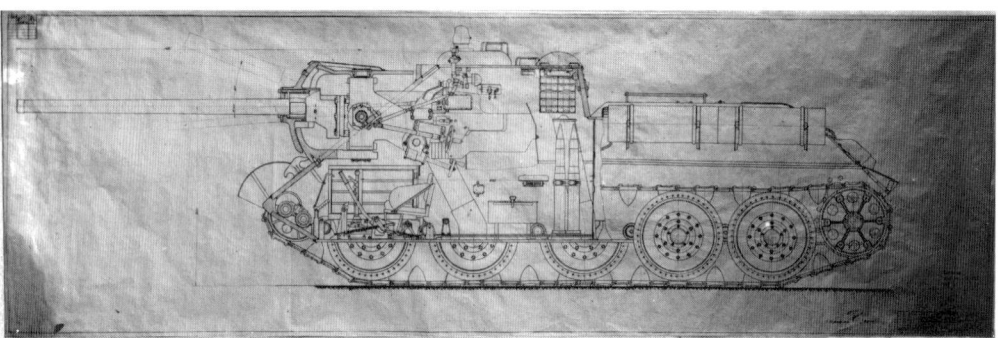

Cutaway right side view of the SU-85-IV. Note the ammunition stowage with 85mm fixed rounds. Introduction of the fixed rounds was a significant improvement, allowing a better rate of fire and ammunition handling. (*Source: RGAE*)

The SU-85-IV prototype seen from the front right. The size of the gun mantlet is clearly visible. Trials revealed the same problem encountered earlier with the SU-122 SPG, where the bulky gun shield partially restricted field of vision for the driver. (*Source: RGAE*)

The SU-85-IV prototype during the trials. This photo gives a good view on the details of the gun shield and sights. All SU-85 prototypes were equipped with a Hertz and PTK panoramic sights. The commission recommended to eliminate the PTK and replace it with the British Mk-IV sight. (*Source: RGAE*)

The SU-85-IV seen from another angle. The SU-85 SPGs were well-protected for 1943 and had a low profile. The height of the vehicle was only 2m, lower than a SU-76M SPG's height. (*Source: RGAE*)

Due to the poor ergonomics, low reliability of the S-18 gun system and low rate of fire, both the SU-85-I and SU-85-IV prototypes were rejected. Pictured: a rear view of the SU-85-IV prototype at the proving grounds. (*Source: RGAE*)

SU-85 Serial

An SU-85 from the first serial production batch at the UZTM factory, apparently after a test run. Note the muddy suspension elements. The most distinctive features allowing to identify vehicles from the early production batches are long handrails on both sides of the superstructure, pistol ports above the handrails and four bolts on the gun mantlet. (*Source: TsAMO*)

Pictured: an early serial SU-85 at the factory yard. After the comparative trials in July–August 1943, the commission recommended to fix deficiencies in the gun design and running gear, as well as to implement a more spacious fighting compartment. However, there was no time to implement changes and the serial vehicle was almost identical to the SU-85-II prototype, including numerous ergonomic-related issues. (*Source: TsAMO*)

Pictured: the SU-85 with its gun at maximum depression. Note that serially-produced vehicles had a crew of 4 (commander, driver, gunner and loader), while the prototypes had a crew of 5. The SU-85 was accepted into service on August 8, 1943, serial production began the same month. (*Source: RGAE*)

This is the late version of the SU-85 introduced in September 1943. The "eyebrows" above the pistol ports were removed and the hoisting eyes on the gun mantlet were a different design. Another detail that immediately stands out is the add-on armor on the commander's cupola. Note that the gun shield is still attached with four bolts. (*Source: RGAE*)

Early SU-85 self-propelled guns loaded on a train in the fall of 1943 in Sverdlovsk (Yekaterinburg). The photo allows to examine the details of the add-on armor on the commander's cupola from another angle. (*Source: TsAMO*)

In November 1943 the design was again slightly modified. This version has a gun shield fitted with six bolts. The pistol port in the right side was moved backwards and placed under the handrails. The other changes included a different shape of the connecting beam—it became more sharp-angled—and a tracklink attached to the front glacis. (*Source: TsAMO*)

The same late SU-85 seen from the left side. Another visible change was that the long handrails were eliminated and replaced with shorter ones. (*Source: TsAMO*)

The same SU-85 seen from the top-right. This vehicle was produced in February 1944. The new design of the commander's cupola is clearly visible. (*Source: TsAMO*)

SU-85BM-I

The experimental SU-85BM-I SPG was designed from August to December 1943 at UZTM's design bureau. The prototype was delivered in December and tested in January–February 1944. The vehicle was built on the serial SU-85 chassis and armed with an 85mm D-5S-85BM high-power gun with a barrel extended by 1.068m. The design allowed muzzle velocity of 800–900m per second—enough to penetrate 100mm frontal armor of the German Tiger I tank at 1500m. The prototype passed factory and field tests, bat was rejected in favor of another prototype armed with a 100mm D-10S gun. Pictured: SU-85BM-I SPG seen from the right side during trials in winter of 1944. (*Source: RGAE*)

SU-85BM-II

The SU-85BM-II was an experimental vehicle based on the SU-100 medium self-propelled gun chassis. The prototype was armed with a 85mm D-5S-85BM high-power gun developed by Factory's #9 design bureau. This variant has a barrel extension that allowed muzzle velocity of 1050m per second and was able to penetrate frontal armor of the German Tiger I tank at 2000m. (*Source: RGAE*)

The same SU-85BM-II seen from the right side. The D-5S-85BM had a maximum elevation of +26 degrees and depression of -3 degrees. (*Source: RGAE*)

The same SU-85BM-II with its gun in travel position. While the firepower increased significantly, the longer gun barrel had its disadvantages, making it difficult to navigate the SPG through highly congested terrain such as cities or dense forests. (*Source: RGAE*)

SU-85M

In July 1944 UZTM was ordered to start production of the SU-100 SPGs in September. In order to maintain stable production output, UZTM put into serial production a hybrid vehicle designated as SU-85M. Basically, it was an SU-100 armed with a D-5S-85A gun. The first serial SU-85M was delivered in July 1944 and remained in production until November 1944. Pictured: a front view of the SU-85M. (*Source: RGAE*)

The SU-85M was well protected, with 75mm-thick frontal armor. Other changes included a commander's cupola, new improved ventilation system and increased ammunition stowage capacity to 60 rounds. The design allowed the 85mm gun to be replaced by a 100mm one without any significant changes in assembly and production technology. (*Source: RGAE*)

The same SU-85M seen from the rear. The photo allows to see the details of the engine deck and rear side of the superstructure and commander's cupola. (*Source: RGAE*)

SU-76i

Since the development and production of the Soviet SPGs lagged behind expectations, it was decided in January 1943 to convert captured Panzer III tanks and StuG III assault guns into an SPG. This step allowed to fill (at least partially) the critical capability gap on short notice and rapidly equip the Red Army with light self-propelled guns. Pictured: a right side view of the serial SU-76i. (*Source: RGAE*)

The SPG was designed in January–March 1943 at the Factory #37 in cooperation with TsAKB. The prototype was tested in March and the serial vehicle was designated SU-76i. The I stands for "Inostranniy" or foreign. Interestingly, the SU-76i was accepted into service in January 1943, long before the prototype was delivered and finished acceptance trials. The photo shows the same vehicle from another angle. Note the details of the gun mount. (*Source: RGAE*)

The SU-76i received a new armored superstructure and was rearmed with a 76.2mm F-34 tank gun, while the engine, transmission and running gear remained the same as on Panzer III tanks and StuG III assault guns. Pictured: the same vehicle seen from the rear with the armored door open. (*Source: RGAE*)

An interior shot of the SU-76i giving a good view of the gun breech and ammunition stowage. The loader sat behind the gunner, slightly more towards the center, while the commander sat on the right side of the gun. (*Source: RGAE*)

Another interior shot, showing the left part of the fighting compartment with a gunner's seat and an ammunition rack. (*Source: RGAE*)

This photo shows a command version of the SU-76i equipped with a commander's cupola. This version entered serial production in August 1943, the design of the cupola was copied from the German Panzer III tank. A total of 200 SPGs were produced by Factory #37 at Sverdlovsk (Yekaterinburg) between May and late November 1943. While the SU-76i proved itself effective and allowed to reuse captured materiel, it was difficult to use and maintain these vehicles in the frontline units due to the lack of spare parts, qualified technicians and even service manuals. (*Source: TsAMO*)

SU-152 (KV-14)

The KV-14 was the prototype of the heavy self-propelled gun based on the KV heavy tank chassis. The project was initiated in 1942, the prototype was delivered on January 24th, 1943 and designated KV-14 (Object 236). Pictured: a front view of the KV-14. (*Source: RGAE*)

The KV-14 prototype seen from the rear. The main purpose of the KV-14 was to engage and destroy enemy fortifications with direct fire. However, sometime it was used in an anti-tank role.

The same prototype seen from the left side. The KV-14 was heavily armored and armed with a 152.4mm ML-20S gun-howitzer. The SPG was able to fire directly and indirectly, but in practice the latter mode was rarely used. The slogan on the side translates as "Death to the German occupiers!"

The KV-14 seen from the right side with its gun at maximum elevation. The KV-14 was tested in early February and accepted into service on February 14, 1943. Note the slogan on the side: "A gift to the Red Army's 25th anniversary." (*Source: RGAE*)

Serial SU-152, March–July 1943 pattern

The photo shows the serial SU-152 produced at the Kirovsky Plant in May 1943. This is what the SU-152 SPGs produced between March and July 1943 looked like. (*Source: TsAMO*)

Since March 1943 the serially produced SU-152 SPGs received some minor improvements, including simplified handrails on the superstructure and simplified design of the cap above the gun mantlet. This particular SU-152 with hull #3046140 was photographed during warranty and performance trials between May and August 1943. (*Source: TsAMO*)

The same SU-152 seen from the front. Despite all efforts, the running gear remained the Achilles' heel of the vehicle. Taking into account the inherent weakness of the Soviet repair and recovery system, it remained a critical issue for Soviet heavy self-propelled artillery throughout the war. (*Source: TsAMO*)

The same SU-152 seen from the rear, probably before the mobility trials. This vehicle failed the warranty trials due to numerous malfunctions in the running gear, including V-2K engine, gearbox, both side clutches, road wheels, tracks and elements of the braking system. According to the warranty, the SU-152 should have been functional for 2000km. However, the tests revealed that it could endure only 400–500km under condition of "thorough technical maintenance," which was, in most cases, impossible to provide. (*Source: TsAMO*)

The same SU-152 seen from another angle. The photo allows to examine the design of the gun mantlet (including the simplified cap) and the roof of the superstructure. (*Source: TsAMO*)

Serial SU-152, late 1943 pattern

The SU-152 climbs up a 30-degree slope. This vehicle was produced at the Kirovsky plant in October 1943 and photographed during the warranty trials in October–November 1943. A number of improvements were introduced to the design in late September 1943. Most of them were minor, but starting with September, the SU-152 received a new ventilation system with two fans installed in the roof of the superstructure. (*Source: TsAMO*)

The same SU-152 goes down a 30-degree slope. The improved vehicles were no more reliable than the earlier batches. This vehicle, hull number 309598, failed the warranty trials due to the same defects in the running gear. Also of note is that the commission pointed to the production defects and low assembly quality, another unresolved issue of the Soviet tank industry. (*Source: TsAMO*)

The same SU-152 moving on a side slope. These vehicles were in service with the Red Army till the end of the Great Patriotic War, a total of 670 SU-152s were produced from February to December 1943. (*Source: TsAMO*)

S-51 SPG (Autumn 1943)

The S-51 was an experimental self-propelled howitzer designed by TsAKB in 1943 under the supervision of V.G. Grabin. (*Source: TsAMO*)

The S-51 was armed with a 203mm howitzer that used the oscillating part of the B-4 gun, but had the ballistics of the B-4BM howitzer. Pictured: the S-51 prototype during factory trials in February 1944, the howitzer is at maximum elevation. (*Source: TsAMO*)

The SPG was based on a KV-1S tank chassis and retained the same arrangement with engine at the back and driver's compartment at the front of the vehicle. This picture gives a good view on the rear part of the howitzer and an engine deck. (*Source: TsAMO*)

The same S-51 seen from the front. While the hull was heavily armored, the 7mm-thick gun shield provided only limited protection against small arms fire and shell splinters. (*Source: TsAMO*)

The S-51 in ready to fire position seen from the right side. The vehicle inherited the same engine and transmission used in the KV-1S tank and was able to move with a maximum speed of 30km/h for a distance of up to 250km. (*Source: TsAMO*)

The S-51 with the howitzer in travel position. Note the barrel recoiled to the back and covered with a tarpaulin. (*Source: TsAMO*)

The S-51 passed factory trials and was even recommended for serial production. However, it was not adopted or serially produced for a number of reasons. Probably the most significant were low stability, and hence low accuracy of the system, and unavailability of the B-4 howitzers which had been taken out of production in 1942. Pictured: the same S-51 in travel position seen from the rear-right. (*Source: TsAMO*)

The same S-51 SPH in transition from travel to combat mode. Note the crew members standing on the unfolded parts of the gun shield and the fully recoiled howitzer. (*Source: TsAMO*)

The S-51 had an enormous crew of 10 and carried 12 shells and primers in wooden crates attached to the engine deck. Since there was no available space for the whole crew, some of them had to travel on foot or on the accompanying vehicles. (*Source: TsAMO*)

SU-11 SPAAG

The SU-11 was an experimental self-propelled anti-aircraft gun developed in October–November 1942 at Factory #38. More significantly, the SU-11 was designed simultaneously with another armored vehicle on the same chassis, the SU-12 assault gun. Essentially it was an attempt to build a family of combat vehicles on a serially produced platform, the T-70B light tank. (*Source: Author*)

The prototype was delivered in November 1942 and in December both vehicles were tested at Gorokhovetsky ANIOP. While the SU-11 and SU-12 successfully passed the gunnery trials, both vehicles failed the automotive tests due to the "imperfect design and manufacturing defects" in the running gear. The SU-11 had a crew of 6 and was armed with a 37mm 61K cannon. It was not adopted or produced, and the single prototype is preserved at Kubinka tank museum near Moscow. Pictured: the SU-11 prototype during the state trials in December 1942. (*Source: TsAMO*)

SU-31

In Spring of 1942 another attempt to build a SPAAG was undertaken by Factory #37, UZTM and engineer S. Ginzburg. Since the design of the SU-31 incorporated components of the T-60 light tank, the prototype was delivered and tested in June 1942. Pictured: a front view of the SU-31 prototype. (*Source: RGAE*)

The same SU-31 prototype with its gun elevated. The SU-31 had a combat weight of 9.5 tons and a crew of 6. It was armed with a 37mm 61K cannon, a DT machine-gun and one PPSh submachine gun. (*Source: RGAE*)

It was possible to use SU-31 in direct fire mode against ground targets. Pictured is the SU-31 with its turret turned to the left and the gun lowered. (*Source: RGAE*)

The SU-31 with its 37mm gun at maximum elevation. The details of the running gear featuring elements of the T-60 light tank are clearly visible. (*Source: RGAE*)

The SU-31 prototype during comparative mobility trials in late September 1942 where it was tested with the T-70 light tank. (*Source: TsAMO*)

The same prototype during mobility trials on a swampy terrain seen from the left side. Note the three-tone camouflage pattern. (*Source: TsAMO*)

The SU-31 overcomes swampy terrain. The prototype was built of mild steel with 15mm thickness all around. (*Source: TsAMO*)

Another shot from the same comparative trials in September 1942. The SU-31 demonstrated good cross-country capabilities, roughly at the same level as the T-70 light tank. Overall, the SU-31 was evaluated positively and the concept was accepted for further development. (*Source: TsAMO*)

SU-37 / ZSU-37

SU-37, Factory #4 testbed

The SU-37 was a testbed developed at Factory #4 in Krasnoyarsk. In September 1943 this facility was tasked to develop a self-propelled AA-gun, as it was the manufacturer of the 37mm 61-K anti-aircraft gun and its naval version the 70-K. (*Source: TsAMO*)

The project offered by Factory #4 was rejected by Artillery Committee of the Main Artillery Directorate (GAU) in December 1943. Nevertheless, the testbed was built and tested in April 1944. The aim of the trials was to prove the possibility to mount a 37mm 61-K gun on a SU-15M SPG chassis. The secondary goal was to compare its performance with the same system mounted on a 52-U-167 gun carriage. Pictured: a SU-37 prototype with a Soviet officer standing in front of the vehicle to demonstrate its height. (*Source: TsAMO*)

The running gear and engine of the SU-15M SPG remained unchanged. The cannon was mounted on a rotating platform, while the gun shield and radio were not installed for the trials. A SU-37 prototype with its cannon at maximum elevation. (*Source: TsAMO*)

The SU-37 with its crew in the fighting compartment. As with most Soviet SPAAGs, the SU-37 was designed to engage both air and ground targets. (*Source: TsAMO*)

The same vehicle on a 15-degree side slope. There are seven Soviet soldiers and officers in the picture, six of them operate the 61-K gun and the officer on the left commands. The non-self-propelled version of the 61-K gun mounted on four-wheeled ZU-7 carriage had a crew of eight. (*Source: TsAMO*)

The picture taken from another angle shows the SU-37 testbed on the same spot. The driver's hatch is open, but it is not known whether the driver is inside. (*Source: TsAMO*)

The SU-37 after the live-fire trials, the soldier cleans the gun barrel. The results of the trials looked promising and it was recommended to re-test the vehicle after installing the radio set, gun shield and a self-stopping rotating mechanism. (*Source: TsAMO*)

ZSU-37-1, Factory #38 design

Another SPAAG prototype was assembled at Factory #38 in Kirov. Unlike the testbed designed at Factory #4, this vehicle was built on a SU-76M chassis and had an open-topped armored turret. (*Source: TsAMO*)

The vehicle is known as ZSU-37 or ZSU-37-1; however, the designation SU-37 was used in the internal correspondence. Pictured: ZSU-37-1 prototype during trials in January–February 1944. (*Source: TsAMO*)

A front view of the ZSU-37-1 during trials at the NIBT proving grounds. The 37mm 61-K gun is at maximum elevation. Note the radio antenna on the left side of the turret. The vehicle was equipped with a 12 RT-3 radio set and had a TPU-F3R internal communication system. (*Source: TsAMO*)

The ZSU-37-1 seen from the rear. Note the slightly curved rear side of the turret with no doors or hatches. (*Source: TsAMO*)

The ZSU-37-1 had a crew of six. The driver sat in the front section of the hull, while two gunners, a gun-layer, a loader and a commander were located in an open-topped turret. The commander also acted as a left gun-layer. The ammunition was stored in four boxes in the right side of the fighting compartment. The trials revealed severe (but typical for Soviet armored vehicles) ergonomic deficiencies, such as cramped fighting compartment, obstructed access to ammunition boxes and others. (*Source: TsAMO*)

The ZSU-37-1 moves through a snow-covered field during mobility trials. (*Source: TsAMO*)

The ZSU-37-1 prototype climbs up a steep slope during mobility trials in January–February 1944. (*Source: TsAMO*)

The same slope, the ZSU-37-1 moves downhill. The prototype passed the mobility trials, but live fire trials were not conducted, since there were no armor piercing rounds available for testing, the crew was unexperienced and the ammunition stowage was "located inconveniently." (*Source: TsAMO*)

The ZSU-31-1 moves along the 20-degree steep slope at the same location. (*Source: TsAMO*)

The ZSU-31-1 moves over a small hill on the snow-covered field. The SU-37-1 demonstrated approximately the same level of on-road and off-road mobility as the SU-76M. The average speed on road was 24km/h, off-road 15.3km/h, while the average speed on a snow covered field was 12.9km/h. Eventually, the ZSU-37-1 passed all trials and was recommended for serial production. However, Factory #38 had no equipment for production and the project was once again transferred, this time to Factory #40 in Mytischi near Moscow. (*Source: TsAMO*)

ZSU-37-2, Factory #40 design

The second prototype (SU-37-2) was delivered by Factory #40 in spring of 1944 and tested in July. While the SU-37-2 was also built on the SU-76M chassis, the engineers of the Factory #40 made some modifications to the design. Pictured: the SU-37-2 prototype apparently during trials in July 1944. New design of the turret, arrangement of ammunition stowage boxes, tarpaulin covers, modified rear hull and other changes are clearly visible. (*Source: TsAMO*)

While the vehicle was modified, it was still designated as a "SU-37 based on a serial SU-76M" in the official documentation. Here is the second prototype seen from the left side with the 61-K autocannon elevated at an angle of 80 degrees. (*Source: TsAMO*)

An interesting solution was chosen to simplify access to the ammunition stowage. The ammunition boxes were fitted to the rear side of the turret in two rows; it was possible to flip boxes in the upper row to access ammunition in the lower row. (*Source: TsAMO*)

The SU-37-2 from the right side with its cannon at its maximum depression of 2 degrees. (*Source: TsAMO*)

A front view of the SU-37-2, the shot was taken during the summer trials. The vehicle has noticeable differences, such as missing boxes for spare parts and tools on the left and right track fenders, as well as there is no radio antenna on the turret. While the prototype was equipped with a 12-RK radio set, according to the official documentation, it is possible that the radio apparatus was fitted after the trials. (*Source: TsAMO*)

The same SU-37-2 seen from the rear. The prototype passed live fire and mobility trials and was recommended for serial production. However, the commission recommended some minor deficiencies, mostly regarding ergonomics, be rectified. (*Source: TsAMO*)

ZSU-37-3, Factory #40, the third prototype

The final prototype or SU-37-3 was delivered and tested in October–November 1944 at NIBT proving grounds. The SU-37-3 passed the tests and entered serial production in 1945. The vehicle had very few changes compared to the SU-76M or earlier prototypes, however a number of minor improvements were made, based on the results of the SU-37-2 trials. (*Source: TsAMO*)

An overhead view of the vehicle's fighting compartment. The SU-37-3 received a shortened spent-case ejection chute, among other minor improvements. (*Source: TsAMO*)

The SU-37-3 in travelling position. The fighting compartment and gun barrel are covered with a tarpaulin. (*Source: TsAMO*)

The same prototype seen from the right side, the rectangular box on the side of the hull is an air inlet cover. While the trials were a success, the serial ZSU-37 was a classic example of the "too little too late" result. ZSU-37 was the only Soviet self-propelled anti-air gun that went into serial production during wartime, however only 12 vehicles were produced in February–April 1945. A total of 75 ZSU-37s were produced between 1945 and 1948. (*Source: TsAMO*)

T-70-based 37mm AA gun (ZUT-37 early)

In 1942–43 the Soviet command was determined to increase the air defense capabilities of the Red Army's tank forces and initiated several development projects of self-propelled anti-air guns based on serially produced light tanks. Pictured above is an experimental vehicle tested at GABTU's proving grounds in September–October 1942. Essentially it was a T-70 light tank armed with a 37mm Shpitalny cannon and a DT machine gun in a heavily modified turret. According to documentation, the vehicle was designed as a dual-purpose combat vehicle able to engage both air and ground targets. (*Source: TsAMO*)

Initially the turret had an armored roof, but at some point it was replaced with an open-topped version. Trials showed numerous design and mechanical flaws, including low accuracy of the cannon and unsatisfactory rate of fire. On top of that, the crew arrangement predetermined the failure of the whole concept. Since the experimental SPAAG had a crew of two, the commander had to act as a gunner, commander and loader, while being in combat. The commission recommended to fix the conceptual and design flaws, but eventually the project was abandoned. (*Source: TsAMO*)

T-70-based twin MG mount

Another SPAAG testbed was armed with a twin DShKT machine-gun mount in an open-topped turret. Note that the engineers used an original turret from the serial T-70 light tank. (*Source: TsAMO*)

The experimental vehicle was tested in December 1942. The trials showed multiple critical issues, including unsatisfactory accuracy, difficulties with aiming and insufficient ammunition surplus, to name a few. Also the design suffered from typical ergonomic-related problems. On top of that, the prototype was not equipped with a radio, an obvious disadvantage for a vehicle intended for air defense of fast-moving mechanized units. As a result, the design was rejected. Pictured: an overhead view of the twin MG mount in the turret. (*Source: TsAMO*)

ZUT-37 SPAAG (late)

The ZUT-37 was another attempt to build a SPAAG armed with a 37mm cannon on a T-70 chassis, in fact a further development of the experimental vehicle tested in September–October 1942. The new variant was developed in OKB-15 design bureau and tested in December 1942–January 1943. The engineers fixed some deficiencies revealed during the trials in the autumn of 1942 and offered a completely redesigned turret. Pictured: the ZUT-37 SPAAG with its gun in travelling position. The vehicle was based on a serial T-70 tank (chassis #4204017) with a new experimental ZUT-37 turret. (*Source: TsAMO*)

The same vehicle with its gun elevated at an angle of 45 degrees. The ZUT-37 was tested in December 1942–January 1943 with discouraging results. While the cannon worked relatively well and was reliable, the gun mount was unbalanced and caused low accuracy and precision during the live-fire test. (*Source: TsAMO*)

The vehicle was armed with the same 37mm Shpitalny cannon. Accordingly, the ammunition feed and storage boxes were adopted for 37mm rounds. Pictured: the ZUT-37 prototype with its gun at maximum elevation of 77 degrees. The turret was protected with 16mm-thick armor all round. (*Source: TsAMO*)

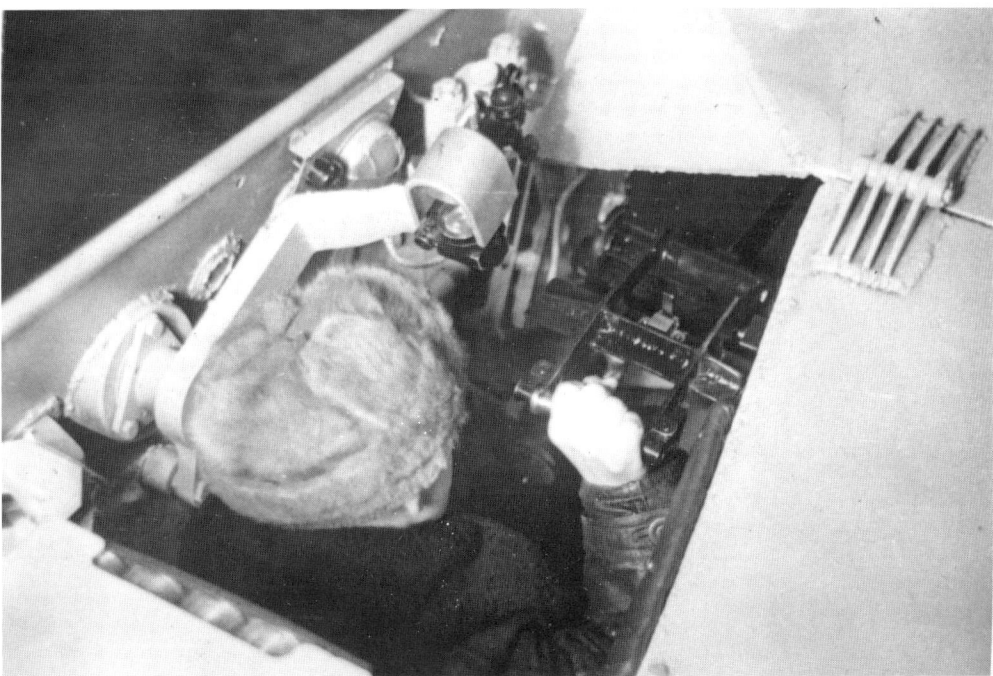

The ZUT-37 had a crew of two that consisted of a driver and a commander, who also acted as a loader and gunner. Pictured: a Soviet soldier opens an ammunition feeder to reload the autocannon. Note his uncomfortable position and the obviously cramped interior of the turret with the massive breech of the cannon to the right of his seat. Eventually the ZUT-37 project was halted along with other self-propelled anti-aircraft gun projects on serial light tank chassis. The only exception was the T-90-based SPAAG that passed the trials, but was not put into serial production. (*Source: TsAMO*)

1944–1945

SU-85B

The SU-85B was an improved variant of the SU-85A. The prototype was delivered in early 1945 and tested in April–May. An upgraded version received an 85mm LV-2 gun with a muzzle brake, more powerful GAZ-80 power plant and thicker armor. (*Source: TsAMO*)

A right-side view of the modified SU-85B. The vehicle is seen in the factory yard, apparently before or right after the factory trials. (*Source: TsAMO*)

A front view of the same SU-85B, allowing to see the details of the muzzle brake and gun mantlet. (*Source: TsAMO*)

The same vehicle seen from the rear. Note the ammo stowage attached to the rear door. The SU-85B passed state trials in April–May 1945 and was recommended for serial production. However, it was never adopted into service or produced. (*Source: TsAMO*)

GAZ-75

The GAZ-75 (factory index) was developed in 1943 at GAZ Automobile Plant's design bureau, the prototype being delivered and tested in 1944. The vehicle was also known as a "SU-85 self-propelled gun based on T-70 tank" or SU-85 GAZ. In fact, the GAZ-75 was the further evolution of the SU-74D experimental SPG. Components of the SU-76M SPG and T-70 light tank were used in the design. (*Source: TsAMO*)

The SPG was armed with a 85mm D-5S-85A gun. In comparison with other designs the GAZ-75 had the thickest frontal armor of 82mm (frontal glacis plate), while SU-76M had only 25mm and UZTM's SU-85 had 45mm thick frontal armor. On the other hand, the SU-85 had 45mm thick side and rear armor, while the GAZ-75 protection was significantly lower with only 15mm thick side and rear armor. Pictured: the GAZ-75 prototype seen from the left during the trials in January 1944. Note the rectangular hatch for loading ammunition in the left side of the superstructure. (*Source: TsAMO*)

A front view of the GAZ-75 prototype. The vehicle had a crew of four. The driver and the commander sat to the right of the gun, and the gunner and loader to the left. Note the driver in his hatch and details of the radio antenna mount on the right side. (*Source: TsAMO*)

Rear top view of the prototype with all hatches closed. The commander had a rotating armored cupola with a Mk IV (Vickers) periscopic sight. The driver also had a periscopic sight in his own hatch on the roof. (*Source: TsAMO*)

Rear top view of the same prototype with all hatches open. The GAZ-75 was equipped with a GAZ 75B petrol engine developing 145 hp. It was found during trials that the vehicle was underpowered, which was not surprising, given that the GAZ-75 had the same power output as SU-76M, but was 34% heavier (14.1 tons versus 10.1 tons). (*Source: TsAMO*)

The GAZ-75 moves on a country road during mobility trials. The vehicle was able to move with an average speed of 30.2km/h on a paved road and 17.2km/h on a country road, while the average off-road speed was only 8.3km/h. (*Source: TsAMO*)

The GAZ-75 tries to climb up a 30-degree slope. The attempt was unsuccessful, trials showing that the prototype was only able to overcome a 24-degree slope in winter conditions. (*Source: TsAMO*)

The same prototype tries to move on a 22-degree side slope. The photo shows the moment when the vehicle starts to slide down the slope. (*Source: TsAMO*)

Pictured: the GAZ-75 during the water-obstacle crossing test in January 1944. The prototype failed to climb up the ice-covered shore of the shallow river. (*Source: TsAMO*)

Another shot of the GAZ-75 crossing a river, this attempt was successful. The test showed that the SPH was able to ford a 0.8m deep water obstacle, but the waterproofing of the bottom needed improvement. The GAZ-75 prototype passed the trials, but the project was discontinued. While the design had no critical flaws, it was not significantly better in comparison to the SU-76M or SU-85. (*Source: TsAMO*)

SU-57

The SU-57 was an experimental self-propelled anti-tank gun based on the robust SU-76M platform. The idea arose in 1943 and the prototype was delivered by spring of 1944. (*Source: TsAMO*)

The first prototype of the SU-57 SPG had an open-topped fighting compartment. This variant was tested in May 1944 at Gorokhovetsky ANIOP. (*Source: TsAMO*)

The SU-76M (SU-15) was taken as a basis for the prototype without any significant changes in the hull, running gear and internal arrangement. The gun barrel of the 76mm ZiS-3 gun was removed and replaced with a 57mm barrel of the ZiS-2 system. However, the optics of the gun remained the same. Pictured: the same prototype seen from the rear. Note the serial number 404855 painted to the left of the rear door. (*Source: TsAMO*)

Following the initial tests, the prototype was improved. The second variant of the SU-57 SPG had a fully enclosed fighting compartment and a new 10-T telescopic sight. Pictured: the SU-57 second variant during trials in June 1944. The serial number indicates that this is the same vehicle. (*Source: TsAMO*)

The SU-57 second prototype seen from the rear. Note the Mk IV periscopic sight on the roof of the fighting compartment. The second prototype passed the trials and was greenlighted for serial production. It was planned to start production in October 1944 at the GAZ plant. However, this plan was never implemented and the SU-57 remained a prototype. (*Source: TsAMO*)

SU-122P

The SU-122P appeared as a result of the further development of the SU-100 SPG. The project was finished in May 1944, while the prototype was built in September. The SU-122P successfully passed the trials, but was not adopted and put into serial production. (*Source: TsAMO*)

The SU-122P was armed with the D-25S gun. The powerful 122mm rifled gun was mounted with the same components as a D-10S gun, allowing to put the new model into serial production in an expedited timeline. Pictured: the SU-122P prototype seen from the right side. The gun is at maximum elevation of 20 degrees. (*Source: TsAMO*)

The same SU-122P photographed from the left. The gun is at maximum depression of -2 degrees. The shot was taken during the trials in November 1944 at NIBT proving grounds. (*Source: TsAMO*)

A rear view of the SU-122P experimental SPG allowing details of the hull and superstructure to be seen. (*Source: TsAMO*)

The SU-122P was built on the chassis of the serial SU-100 SPG produced at the Uralmash plant. This shot allows the roof of the superstructure and details of the engine deck to be seen. (*Source: TsAMO*)

Note that exterior of the SU-122P was almost identical to the SU-100. The most prominent external difference was the longer (3.7m) gun barrel and muzzle brake. Pictured: the SU-122P prototype tries to climb the 34–36-degree slope. This attempt was unsuccessful. (*Source: TsAMO*)

The SU-122P prototype successfully overcomes the same slope. (*Source: TsAMO*)

The SU-122P prototype moves down the slope. The trials showed that the length of the gun barrel could cause problems. The gun's muzzle touched the ground while the vehicle moved up or down steep hills, or on bumpy roads with 1–1.2m deep potholes. (*Source: TsAMO*)

This shot illustrates the aforementioned problem. The SU-122P touches the ground with its gun muzzle while moving through a large pothole on a forest road. (*Source: TsAMO*)

The SU-122P on a 25-degree side slope. The vehicle shows no signs of slipping. (*Source: TsAMO*)

While the SU-122P proved able to ford water obstacles up to 1m deep, but the risk of scooping up water with the gun was extremely high. The recommendation issued by the trial commission was to cross water obstacles with a slight tilt to the left side. It is unclear how the Soviet tankers were supposed to follow this recommendation in a combat environment, but it was not an isolated event, since Soviet industry tried to address design flaws by issuing strict rules of use for the materiel. (*Source: TsAMO*)

The SU-122P fords a 1m-deep water obstacle. While the prototype passed mobility and gunnery trials, it was not accepted into service or produced serially. There were several reasons for rejecting the SU-122P. Firstly, there were other SPGs armed with 122mm guns. Secondly, 122mm guns were needed for IS-2 heavy tank production, therefore these weapons were in short supply. Thirdly, the medium SPG chassis offered limited space for the crew allowing to place only 26 rounds and propellant charges in the fighting compartment. Moreover, cramped fighting compartment caused a low rate of fire of only 2.1–2.75 rounds per minute, which was deemed insufficient. (*Source: TsAMO*)

KSP-76 (GAZ-68)

The KSP-76 was one of the few attempts to build a self-propelled gun on a wheeled chassis during the war. The prototype was developed at GAZ plant's design bureau in 1943, while the prototype was delivered in 1944. Pictured: the KSP-76 prototype during the trials at NIBT proving grounds in September 1944. (*Source: TsAMO*)

The KSP-76 was a wheeled SPG intended for accompanying infantry and cavalry units and providing them with direct fire support. The name "KSP-76" stands for "Kolyosnaya Samokhodnaya Pushka" or Wheeled Self-Propelled Gun with a 76mm gun. The vehicle was based on the experimental GAZ-63 4x4 truck chassis. Pictured: the KSP-76 prototype seen from the right side with its gun in travel position. (*Source: TsAMO*)

A rear view of the KSP-76 and engine louvres. The GAZ-202 engine, output 85 hp, was mounted in the rear compartment. The fighting compartment in the middle was open-topped and lightly armored. (*Source: TsAMO*)

One of the undeniable advantages of the vehicle was its low profile. The KSP-76 was only 55cm tall, 54cm shorter than SU-76M. Another strong point was its superb mobility on paved and country roads. However, the off-road mobility of the KSP-76 was unsurprisingly worse than the SU-76. Pictured: the same prototype with its gun elevated at an angle of 15 degrees. (*Source: TsAMO*)

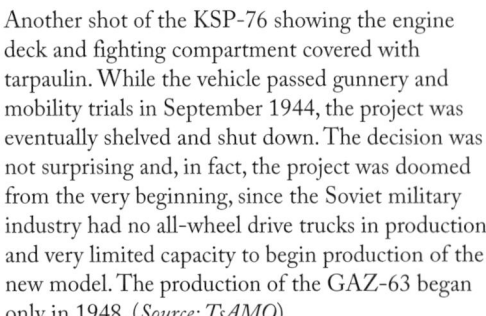

Another shot of the KSP-76 showing the engine deck and fighting compartment covered with tarpaulin. While the vehicle passed gunnery and mobility trials in September 1944, the project was eventually shelved and shut down. The decision was not surprising and, in fact, the project was doomed from the very beginning, since the Soviet military industry had no all-wheel drive trucks in production and very limited capacity to begin production of the new model. The production of the GAZ-63 began only in 1948. (*Source: TsAMO*)

OSU-76

The OSU-76 was another experimental light self-propelled gun intended for direct infantry support. The OSU-76 was developed in March–May 1944 at the Factory #38 design bureau. Three prototypes were delivered and tested in the summer of the same year. Pictured: the OSU-76 prototype seen from the left side. The 76mm gun is in travel position (not the travel lock on) and the fighting compartment is covered with tarpaulin. (*Source: TsAMO*)

The same prototype photographed from the top-right with the tarpaulin removed. The OSU-76 had a crew of three, the driver, the gunner-commander and the loader. All three crewmembers sat in the open-topped fighting compartment. The driver's place was to the right of the gun, while the commander and loader's stations were positioned on the other side of the gun. (*Source: TsAMO*)

The right-side view of the OSU-76 with the gun at maximum elevation of 12 degrees. While this SPG belonged to the same class as SU-76M and was armed with the same 76mm ZiS-3 gun, it was significantly smaller and had a combat weight of only 4.18 tons. In comparison, the SU-76 weighed 10.5 tons. (*Source: TsAMO*)

The same prototype with the gun at maximum depression of -4 degrees. The OSU-76 was a small and easy-to-conceal vehicle, it was only 1.56m high, 2.15m wide and 3.9m long. Note the number "5" painted on the side of the superstructure. Another prototype had number "3", the number of the third vehicle is unknown. (*Source: TsAMO*)

The OSU-76 was powered with a four-cylinder GAZ-M1 engine developing 50 hp. The running gear and components of the transmission were taken from the T-60 light tank. Pictured: a front view of the OSU-76 with its gun at maximum elevation. Note the large rectangular air inlet grille on the front-right part of the hull. Clearly not a wise engineering solution for an infantry close support vehicle. (*Source: TsAMO*)

A rear view of the same prototype showing the details on the fighting compartment. Note the open rectangular hatch in front of the driver's seat. (*Source: TsAMO*)

A closer view of the fighting compartment. The commander's seat is visible to the left of the gun breech. The loader's seat was fitted to the left wall of the fighting compartment behind the commander's seat. Due to its size, the OSU-76 had limited space for crew and ammunition resulting in poor ergonomics. (*Source: TsAMO*)

Pictured: another photo, showing details of the driver's station. The result of the trials was not conclusive. The OSU-76 was lighter, simpler and significantly cheaper in production. For example, it required 2.5 times less steel compared to the SU-76M. Additionally, the lightweight OSU-76 showed better off-road performance than a SU-76M. On the other hand, the OSU-76 had poor ergonomics, limited traverse angle and unsatisfactory stability when firing. While the commission recommended the OSU-76 for serial production, it was never adopted or produced. (*Source: TsAMO*)

SU-100 (Object 138)

SU-100, The first prototype

The first prototype of the SU-100 (left) standing next to the serial SU-85 (right). The photo was taken in February 1944. The SU-100 can be easily distinguished by the commander's cupola and larger gun barrel. (*Source: RGAE*)

The SU-100 was developed in 1944 at Urarmash (UZTM), and the D10-S was developed by Factory #9. The first prototype was delivered in February 1944 and entered trials in March 1944. Pictured: the first prototype during trials at Gorokhovetsky proving grounds in March 1944. (*Source: TsAMO*)

The SU-100 was designed as a tank hunter able to deal with the most heavily armored German panzers and assault guns, such as the notorious Panzerkampfwagen VI Tiger and Panzerkampfwagen V Panther tanks. However, SU-100 SPGs were occasionally used for direct infantry support, attacking field fortifications in urban fights and in other roles and combat environments. Pictured: the first prototype photographed from the right side. Note the commander's cupola situated approximately in the middle of the right-side armor plate of the superstructure. In the later variants the cupola was moved backwards. (*Source: TsAMO*)

The first prototype during mobility trials. The vehicle travelled for 564km, 439 of them on snow-covered roads and additional 125km off-road. More to that, the prototype covered another 300km during gunnery trials. While some minor deficiencies were identified, the final conclusion was positive. The SU-100 prototype survived trials and was recommended for serial production with some improvements. (*Source: TsAMO*)

SU-100, *the second prototype*

Following the trials of the first prototype, the second improved version was delivered in May–June 1944. In order to improve working conditions, the commander's cupola was moved to the right, while armor of its front part was thickened to 90mm. Other changes included new ventilation fans, new TSh-15 sight and some other minor modifications. While efforts to fix ergonomic issues were made, the result was far from ideal. Trial commission stressed that the gun crew (the commander, the gunner and the loader) had only 1.2m² of space to work in, and described conditions as "cramped and uncomfortable." The second prototype (pictured) with minor changes was tested in June 1944 and was accepted for serial production, which began in September 1944. (*Source: TsAMO*)

The same prototype seen from the left side. The SU-100 had a crew of four, the driver, the gunner and the loader sat on the left side of the fighting compartment. One of the ammunition racks was also located on the left side of the superstructure. The commander's station was situated to the right of the gun, in the small sponson. SU-100 was in serial production from September 1944 till January 1, 1948. In total, Soviet industry produced 1560 SU-100 SPGs by June 1, 1945 and 2335 vehicles by October 1, 1945. (*Source: TsAMO*)

While the SU-100 was arguably one of the best Allied medium SPGs and probably the most successful Soviet tank destroyer developed during the Great Patriotic War, it was adopted too late and appeared on the battlefield only in January 1945. Ironically, the SU-100, designed as a means against heavy armored German panzers, saw wider service during the Cold War. Pictured: the second prototype seen from the rear. From this side the SU-100 was almost identical to the late production vehicles. (*Source: TsAMO*)

SU-101 and SU-102

The SU-101 (Uralmash-1) was designed at UZTM in the fall of 1944, the project supervised by L.I. Gorlitsky. Two prototypes were delivered by May 1945 and tested in the summer-fall of 1945. While the results of the trials were mostly positive, neither SU-101, nor SU-102 were adopted or produced serially. Both vehicles were originally designed as medium tank destroyers, but eventually the experimental vehicles were used as testbeds to examine the possibility of increasing firepower by installing large caliber guns on a medium chassis. (*Source: TsAMO*)

The SU-101 and SU-102 had almost identical hulls, the design also incorporated components of the T-44 tank. The most prominent distinctive feature was the armament. While the Uralmash-1 or SU-101 was armed with a 100mm D-10SK gun, the SU-102 was armed with a 122mm D-25S gun. Due to weight and size of the gun and ammunition, the SU-102 was 15m longer (7m versus 6.85m) and a bit heavier (34,774kg versus 34,087kg) than the SU-101. Pictured: the SU-101 prototype crosses a water obstacle during mobility trials. (*Source: TsAMO*)

There were two attempts to cross the river. The first try was unsuccessful, and water got into the fighting compartment due to the hatch in the belly was not shut tightly. After this problem was solved, the prototype successfully traversed the river. (*Source: TsAMO*)

Another shot of the river crossing. The prototype was able to ford water obstacles 0.9m deep and 26m wide, as well as to negotiate 30-degree slopes. (*Source: TsAMO*)

The SU-101 crosses the river with crew and engineers sitting on top of the vehicle. We can see five or six men plus the driver not visible on the photo, while both prototypes had a crew of four. Note the rectangular rear door, resembling doors of modern self-propelled howitzers. The door was intended to load ammunition during indirect fire missions, as both prototypes were (at least theoretically) able to fire indirectly. Mounting brackets to the left of the door were used for external fuel tanks. (*Source: TsAMO*)

Another mobility test was conducted on a rainy day. The SU-101 easily climbed up the 23-degree and 100m long wet slope and moved down the same slope. No sliding or other issues were identified. (*Source: TsAMO*)

The SU-101 moves uphill during the same mobility test. The trials commission noted that the 23-degree slope was not, probably, the maximum slope the vehicle was able to overcome. However, they were unable to find a steeper slope on proving grounds and test their theory. (*Source: TsAMO*)

Also, the SU-101 was able to move along and turn on the 23-degree side slope. Again, no issues were identified.

The SU-101 demonstrated roughly the same level of mobility as SU-100 with average speed of 25km/h and maximum speed of 50km/h. The main problem revealed during factory trials was overheating of the engine, gearbox, and poor working conditions for the crew, especially for the driver. Initially there was no bulkhead between the engine and driver's compartment. As a result, the driver suffered from extreme heat and had to leave the vehicle after driving only 7km during trials. Working conditions for the rest of the crew were also unsatisfactory due to the cramped fighting compartment. Later the prototypes were improved and the double-walled bulkhead with asbestos insulation was fitted between the engine and driver's compartment. (*Source: TsAMO*)

SU-100 standing near the Uralmash-1 prototype, probably in the summer 1945. The SU-101 looks significantly bigger, however these vehicles had almost similar dimensions. While the SU-100 was 6.1m long (9.4m including the gun barrel), 3m wide and 2.28m high, the SU-101 was 6.85m long, had a width of 3.1m and overall height of 2.4m. (*Source: TsAMO*)

The SU-102 (Uralmash-2) is easily identifiable by its main gun with a muzzle brake. The 122mm D-25S gun offered better lethality at a cost of less ammunition carried on board. While the SU-101 carried 36 rounds of ammunition, the SU-102 carried only 28. Note two external fuel tanks on the rear plate and a DShK 12.7mm machine gun on a pintle mount. (*Source: RGAE*)

The same SU-102 (Uralmash-2) prototype seen from the left side. Relatively low profile coupled with thick armor offered excellent protection. Both prototypes had 120mm thick frontal armor, while sides of the superstructure were protected with 90mm armor. The SU-101 and SU-102 were, arguably, the most heavily armored SPGs in 1944–45. While these vehicles had strong advantages over existing models and satisfactory passed mobility and live fire trials, neither prototype was found suitable for adoption or at least finalized. (*Source: RGAE*)

ISU-122

ISU-122 (Object 242)

The ISU-122 (Object 242) was developed in 1943 at the Factory's #100 design bureau. The prototype was tested in December 1943 and accepted for serial production, which began in April 1944. The prototype was equipped with two experimental features: wide tracks and a mount for 12.7mm DShK machine gun. Note that the machine gun in this picture is in the anti-aircraft position. Soviet command was concerned about the threat posed by German aircraft since 1942, however the official order allowing to equip SPGs with secondary armament for self and anti-aircraft defense was issued only in October 1944. Pictured: the prototype of Object 242 at the Factory's #100 yard, December 1943. (*Source: TsAMO*)

The SU-122 had the same hull, running gear, engine, optics and electrical equipment as the ISU-152 heavy SPG. The main difference was the 122mm A-19 field gun and corresponding changes in ammunition stowage and gun mantlet. There were two main reasons for having two different-caliber guns on the same chassis. Firstly, the Soviet command wanted to accelerate production output for heavy self-propelled artillery. The second reason was that the A-19 system offered better armor-piercing capability in comparison with the 152mm ML-20 gun used on ISU-152 SPGs. Pictured: the same prototype, but the machine gun is fitted on an experimental tripod. (*Source: TsAMO*)

The new design allowed to increase ammunition capacity to 30 rounds versus 20 in an ISU-152, and partially improved working conditions for the crew. The latter factor increased the rate of fire to average of 2–3 rpm and maximum of 4–5 rpm with a trained crew. Pictured: an ISU-122 with A-19 gun. Serial production model, spring-summer of 1944. (*Source: TsAMO*)

The ISU-122 was designed as a heavy tank destroyer able to counter heavily armored German panzers that proliferated during 1943–44. Indeed the 122mm A-19 gun offered 20–25% better penetration compared to the 152mm ML-20. Trials showed that the A-19 was able to penetrate a 110mm-thick armor at a distance of 2000m. Pictured: left side view of a serially produced ISU-122. (*Source: TsAMO*)

While the ISU-122 was acknowledged as a superior design to the ISU-152, the Main Artillery Directorate (GAU) pushed for further improvements. In particular, they wanted a new main gun with better automation and a further increase in rate of fire. These requirements were formalized by spring 1944 and in April 1944 NKTP ordered to modernize the ISU-122. Pictured: the same ISU-122 seen from the rear-right. (*Source: TsAMO*)

ISU-122BM (Object 243, ISU-122-1)

The Object 243, also known as ISU-122BM or ISU-122-1, was an initiative development of the Factory #100 design bureau. The vehicle was developed in June 1944 by Factory's #100 design bureau. The prototype was delivered and tested in July 1944. (*Source: TsAMO*)

The ISU-122-1 was based on an ISU-122, the A-19 gun being replaced with another system—the 122mm BL-9 gun. Another visual difference was the slightly changed design of the gun mantlet. (*Source: TsAMO*)

A front view of the ISU-122-1 at the factory yard, probably in June or July 1944. (*Source: TsAMO*)

Due to the defect of the gun barrel, the ISU-122-1 failed the initial live-fire trials in 1944. Another barrel was delivered in February 1945. This time the prototype successfully passed trials, but by that time the war had nearly ended. As a result, this variant was never accepted for serial production. (*Source: TsAMO*)

ISU-122S (Object 249, ISU-122-2)

The modernized version of the SU-122, the SU-122S or Object 249, was designed in April 1944. The vehicle was tested in June and accepted into service in August 1944. According to the initial plan, the SU-122S should have replaced the previous version, but in reality, both versions were in production until mid-1945. Pictured: an Object 249 in June 1944 at Gorokhovetsky proving grounds. This version could be easily identified by the D-25S gun with a muzzle brake and a redesigned gun mantlet. (*Source: TsAMO*)

The SU-122S was armed with a 122mm D-25S gun. Basically, it was the same system that was used with IS-2 heavy tanks, but customized for self-propelled guns. Pictured: the same ISU-122S during trials seen from the right side. Note an experimental machine-gun mount with a 12.7mm DShK. This DShK mount, with some minor improvements, was later accepted as a standard mount for serially produced ISU-122 SPGs. (*Source: TsAMO*)

A front view of the same ISU-122S allowing to see the details of the gun mantlet. The new design of the mantlet allowed the armor thickness to be increased to 90–100mm. (*Source: TsAMO*)

A rear-right view of the same ISU-122S at the proving grounds in summer of 1944. While all three SPGs—ISU-152, ISU-122 and ISU-122S—were accepted for serial production, the Soviet military leadership was not entirely satisfied with the efficiency and armor-piercing capabilities of the A-19, ML-20S and D-25S. These concerns led to further experimentation with more powerful guns. (*Source: TsAMO*)

ISU-122BM (Object 251, ISU-122-3)

The Object 251 or ISU-122-3 was another experimental self-propelled gun developed at the Factory #100 design bureau. The vehicle was based on an ISU-122S with the original A-19 system replaced with a 122mm S-26-1 system. The S-26-1 had the same ballistics as a BL-9 and had minor differences such as horizontal breech block. Pictured: front view of the ISU-122-3 prototype. Note the new design of the gun mantlet. (*Source: TsAMO*)

The ISU-122-3 was delivered in autumn of 1944 and introduced for trials in November. The prototype failed initial trials due to the low barrel life, poor ergonomics and situational awareness. Another problem was the 59.5-caliber (4.8m) long gun, which severely limited maneuverability and posed a high risk of touching the ground while moving through rough terrain. Improvements were completed only in June 1945, but by that time the project was abandoned and later cancelled. (*Source: TsAMO*)

The ISU-122-3 during trials in November 1944. This variant retained the same internal arrangement as the ISU-122S and had a crew of four or five. Ammunition capacity was reduced to 25 rounds. (*Source: TsAMO*)

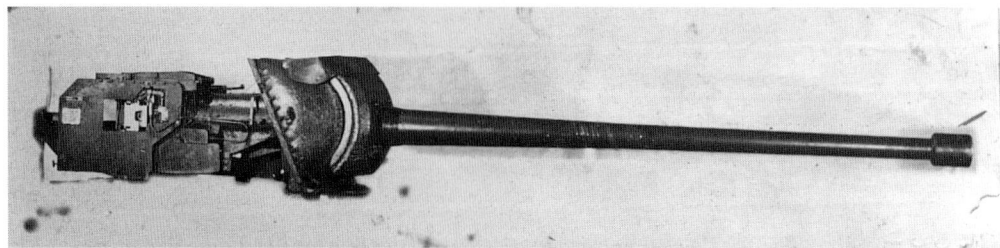

A 122mm S-26-1 system designed by TsAKB. Seen here with a gun mantlet. (*Source: TsAMO*)

ISU-152

ISU-152 (Object 241)

The ISU-152 (other designations, such as IS-152 and Object 241 were also used) was developed from June to September 1943 by the Factory #100 design bureau. The prototype was ready in October 1943 and tested in October and November the same year. Pictured: an early version of an ISU-152 SPG, note the absence of the handrails on the side of the superstructure. This feature was added in later versions. This photo was taken in October 1943 at ChKZ. (*Source: TsAMO*)

The ISU-152 during trials in October or November 1943. This version already has handrails. The trials revealed that the ISU-152 could fire effectively from short stops, while fire on the move was possible, but ineffective. The maximum rate of fire was around 2 rpm. Low rate of fire was typical for Soviet heavy self-propelled guns, but this weakness was not fully resolved until the end of the war. (*Source: TsAMO*)

The serial ISU-152 produced at UZTM in spring or summer of 1944. Note two handrails on the left side of the hull, typical casted front armor plate and fittings for spare tracks on the lower armor plate. In comparison with its predecessor (SU-152), the ISU-152 received more reliable running gear and was better armored. At the same time, it retained a full set of flaws typical for Soviet designs, such as poor ergonomics and inadequate optics and vision devices. (*Source: TsAMO*)

Early variant of the ISU-152. Long handrails on the side of the superstructure suggest that the hull of this vehicle was produced at the Factory #200 between late 1943 and early 1944. (*Source: TsAMO*)

The same early variant ISU-152 seen from the left side. Despite its drawbacks the vehicle was in production from November 1943 and was produced with modifications until 1947. Eventually it became the most produced heavy self-propelled gun with over 2800 units. (*Source: TsAMO*)

ISU-152BM (ISU-152-1, Object 246)

The ISU-152-1 (Object 246) was developed at the Factory #100 design bureau in April 1944 as an initiative. The prototype was delivered and tested in July the same year. The ISU-152-1 was based on the existing ISU-122 design. The main difference was the new 152mm BL-8 (OBM-43) high-power gun and slightly redesigned gun mantlet. The trials revealed serious flaws in the BL-8 gun design and the project was eventually cancelled. The ISU-152-1 was never adopted or accepted for serial production. (*Source: TsAMO*)

ISU-152BM (ISU-152-2, Object 247)

The ISU-152-2 (Object 247) was another pilot design developed by the same design bureau and supervised by Kotin. In essence, the vehicle was an improved Object 246 armed with a 152mm BL-10 gun. Pictured: the ISU-152-2 prototype seen from the left side. The vehicle doesn't have many visual differences from the Object 246 except the BL-10 gun and design of the muzzle brake. (*Source: TsAMO*)

The ISU-152-2 prototype was delivered in August 1944 and tested in December at ANIOP proving grounds near Leningrad. This variant failed the tests for the same reasons as the ISU-152-1 and ISU-122-3. The commission noted low barrel life of the BL-10 system, unsatisfactory traverse angles for the main gun and limited maneuverability due to the 5.38m-long barrel. (*Source: TsAMO*)

A front view of the same ISU-152-2. Note the details of the gun mantlet. The gun is in the travel position. (*Source: TsAMO*)

After trials the BL-10 gun was transferred to Factory #172 for improvement, but this work was not finished until the end of the war. Eventually, the project was cancelled and the ISU-152-2 was never adopted or produced serially. Pictured: the ISU-152-2 prototype seen from the front-left. (*Source: TsAMO*)

ISU-130 (Object 250)

The ISU-130 (Object 250) was an experimental vehicle developed in 1944 at the Factory #100 design bureau, the project was supervised by Josef Kotin. The project emerged as an attempt to further increase the firepower of Soviet self-propelled artillery. The prototype was first tested in November–December 1944. During the winter trials, the vehicle was heavily damaged due to a gun system malfunction. After repairs, the improved prototype was accepted for follow-up trials only in May 1945. The trials took place between June and August the same year. Pictured: the ISU-130 prototype during the trials in the summer of 1945. (*Source: TsAMO*)

The vehicle was based on an ISU-122S SPG and armed with a 130mm S-26 gun developed at TsAKB. The S-26 system was, in turn, developed from the B-13 naval gun and adopted for mounting in a self-propelled gun. Pictured: the same ISU-130 prototype seen from the right side. (*Source: TsAMO*)

As in case with other experimental heavy-caliber self-propelled guns, the ISU-130 retained hull, running gear and electric equipment without any significant changes. The most prominent visual differences were casted gun mantlet, longer gun barrel and square-shaped muzzle brake. Pictured: front view of the ISU-130 prototype during trials in the summer of 1945. (*Source: TsAMO*)

This picture taken from another angle allows details of the S-26 gun and the gun mantlet to be seen. The ISU-130 demonstrated modest results during mobility tests. The average speed while moving on country roads was 12.8km/h, while operational range was only 90km. The latter was recognized as inadequate. (*Source: TsAMO*)

The ISU-130 seen from the right side with its gun in travel position elevated at 14 degrees. Note the 54.7-caliber long gun barrel. This feature, again, became a disadvantage and restricted maneuverability. (*Source: TsAMO*)

While the 130mm S-26 system was deemed superior to the D-25 gun, trials revealed numerous deficiencies of the design including poor operational range, cramped interior and insufficient ammunition load, to name a few. The trials commission emphasized that ergonomic issues caused decrease in combat capabilities so severe that the ISU-130 performed worse than a serially produced ISU-122. Another area of concern was rate of fire. The ISU-130 demonstrated maximum rate of 2.5 rpm (in ideal conditions at proving grounds), but the realistic rate of fire was 1.8–2 rpm that was deemed inadequate. As a result, the ISU-130 was not recommended for serial production. Additionally, the war had ended and Soviet industry switched to other tasks. Pictured: rear view of the ISU-130 prototype. The photo was taken in August 1945 at proving grounds near Kubinka. (*Source: TsAMO*)

Kirovets-2 (Object 704)

The Kirovets-2 (also Object 704 or ISU-152 M1945) was developed in 1945 at the Factory #100 design bureau. The only prototype was built using parts of IS-2 and IS-3 heavy tanks. It was tested in September–November the same year. The vehicle was armed with a 152mm ML-20SM gun-howitzer and two 12.7mm DShK machine guns for self- and anti-aircraft defense. Pictured: front view of the Object 704. The photo allows details of the gun mantlet and machine-gun mount to be seen. (*Source: TsAMO*)

It should be noted that the vehicle was perceived as a "heavily modernized ISU-152," and tested and evaluated accordingly. In comparison with the ISU-152, the Object 704 had a completely redesigned hull with sloped armor. The vehicle was only 2.24m high, but the fighting compartment was wider and generally more spacious. The trials commission emphasized serious improvements in internal arrangement and ergonomics, as well as better organized working conditions for the crew. For example the commanders seat now was located in the direction of travel, while in the ISU-152 the commander sat sideways. (*Source: TsAMO*)

Another improvement was the absence of the muzzle brake, which had partially blocked the field of vision for the gunner in the ISU-122 and ISU-152 SPGs. In addition, the Object 704 featured heavy protection, with 120mm-thick and 50° sloped front upper and lower armor plates, 90mm-thick side plates, and 60–75mm-thick rear plates. In fact, the Object 704 boasted armor protection approximately equal to that of the IS-3 heavy tank. (*Source: TsAMO*)

The Object 704 seen from the right side at NIBT proving grounds in the fall of 1945. The gun is at maximum elevation. The fate of the vehicle was unfortunate. While the Object 704 was an improvement over the existing designs, the GBTU had already developed new requirements for guns and self-propelled artillery, and the project was cancelled. The army expected to receive better SPGs in the future, and some experimental work on new SPGs commenced. Ironically, none of these experimental vehicles was accepted into service or even reached low-rate serial production. (*Source: TsAMO*)

Lend-Lease

M10

The M10 self-propelled gun was the only tracked SPG imported via Lend-Lease. In 1944, the Red Army received 52 of these vehicles. During trials, the M10 SPGs received mostly positive assessments. The commission concluded that M10 vehicles had almost as good mobility as the Soviet SU-122 and SU-85, excellent ergonomics due to the large turret, and good mechanical reliability except for the running gear. The two major flaws of the design highlighted during the trials were the open-topped turret and insufficient obstacle-crossing capability. The M10 SPGs served in two light self-propelled artillery brigades. Since there were no further orders or replacement vehicles, only a handful of them survived until the end of the war. (*Source: TsAMO*)

M15

The M15 self-propelled anti-aircraft gun, officially designated as M15 Combination Gun Motor Carriage, successfully passed the trials at NIBT in 1944 receiving mostly positive evaluation. A total of 100 vehicles of this type were delivered to the USSR where they served as a mobile means of air defense with tank and mechanized troops, as well as fire support SPGs and in reconnaissance battalions. (*Source: TsAMO*)

In 1944, the M15 participated in live-fire test against the German Panzerkampfwagen V Panther tank. The test revealed that the M15's gun was capable of penetrating the 45mm-thick side armor of a Panther at a distance of 200m at 90 degrees using the APC-T M59A1 projectile. (*Source: TsAMO*)

M17

The M17 Multiple Gun Motor Carriage was the most numerous type of self-propelled gun exported through the Lend-Lease agreement, with 1000 vehicles received in 1944–45. After the trials in 1944, the vehicle was immediately recommended for adoption into service with the Red Army. The commission highlighted several advantages, such as good speed and off-road mobility, mechanical reliability, and favorable working conditions for all crew members except the radio operator. (*Source: TsAMO*)

Armed with an M45F quad machine-gun mount, the M17 offered overwhelming firepower against aerial and ground targets, combined with a good on- and off-road mobility and the robust chassis of an M5 half-track vehicle. The M17 proved itself a versatile armored vehicle capable of performing different kinds of missions beyond the intended mobile air defense. (*Source: TsAMO*)

T48

A T48 57mm Gun Motor Carriage was the first self-propelled gun received by the Soviet Union through the Lend-Lease program in 1943. The T48 underwent testing at the NIBT proving grounds in March 1943 and was subsequently adopted into service by the Red Army later that year. In total, 650 vehicles of this type were received during the period 1943–44. These vehicles were known as SU-57s in Soviet service. (*Source: TsAMO*)

The T48 (SU-57) vehicles were designated as light self-propelled guns and were intended for close fire support, similar to the tactical role of SU-76M SPGs. However, the unique combination of firepower and mobility allowed these vehicles to be employed in a wide variety of roles, from reconnaissance and screening missions to deep strikes into the enemy's operational depth. Pictured: a T48 overcoming a steep slope during the trials in the spring of 1943. The vehicle failed to climb up the 22 and 25-degree slopes in winter conditions. The trials revealed vehicle's mobility limitations in cold weather conditions due to the icing of the running gear. (*Source: TsAMO*)

T48 SPGs served in light self-propelled artillery brigades until the end of the Great Patriotic War. The vehicles were highly appreciated by Soviet mechanized troops and participated in many major offensive operations, such as Operation Bagration, and the Berlin and Prague operations. Pictured: the T48 self-propelled gun preserved at the Patriot Park Museum Complex near Moscow. (*Source: Author's collection*)

Notes

Chapter 1

1. Lashkov, A. (n.d.). Ратные дела полковника Василия Тарновского [Military Deeds of Colonel Vasily Tarnovsky]. Military Encyclopedia. https://encyclopedia.mil.ru/encyclopedia/history/more.htm?id=12061911@cmsArticle
2. Ibid.
3. Воеспец. (n.d.). Военный энциклопедический словарь [Military Encyclopedical Dictionary]. Retrieved from https://encyclopedia.mil.ru/encyclopedia/dictionary/details.htm?id=4469@morfDictionary
4. Бехин, И. Б. (1958). Военная реформа в СССР (1924-1925) [Military Reform in the USSR (1924-1925)]. Москва [Moscow]. С. 57 [Page 57].
5. August 1, 1929. Minute #29, "On the system of tank-tractor-auto-armored weapons of RKKA." RGVA, F. 31811, O. 1, D. 7, Pages. 1–2 s ob.
6. June 5, 1929. Triandafillov, V.K. "On the system of tank-tractor-auto-armored weapons of RKKA." RGVA, F. 4, O. 2, D. 504, Pages 5–18 s ob.
7. Ibid. Page 13.
8. Tarasov, A. (2021), *Red Army Auxiliary Armoured Vehicles 1930-1945.* Pen & Sword. Page 56.
9. April 1, 1933. Efimov, N. A. The report of the Chief of the Main Artillery Directorate of the Red Army (RKKA) to the Revolutionary Military Council of the USSR on the revision of the artillery armament system. RGVA, F. 4, O. 14, D. 958, pp. 27–35.
10. August 10, 1933. Voroshilov, K. E. The report of the People's Commissar of the USSR for Military and Naval Affairs and Chairman of the Revolutionary Military Council of the USSR to the Chairman of the Defense Commission of the USSR, V. M. Molotov, on the results of the development of artillery armaments from October 1, 1928, to May 1, 1933, and the prospects for the Second Five-Year Plan. GA RF, F. P-8418, O. 9, D. 15, Pages 65–71.
11. Tarasov, A. (2021), *Red Army Auxiliary Armoured Vehicles 1930-1945.* Pen & Sword. Pages 17–19.
12. May 27, 1940. Kulik, G. I. Memorandum from the Deputy People's Commissar of Defense of the USSR to I. V. Stalin and the Secretary of the Central Committee of the All-Union Communist Party (Bolsheviks) on the absence of self-propelled artillery in the system of armaments of the Red Army and the necessity of adopting four types of self-propelled guns, with attachments. RGANI (Russian State Archive of Socio-Political History), F. 3, O. 46, D. 378, Pages 1–2, 3–33
13. Resolution 219cc. On the organization of the production of 2000 armored tractors. RGASPI (Russian State Archive of Socio-Political History), F. 644, O. 2, D. 6, Pages 153–157.
14. Solyankin, A. G., Pavlov, M. V., Pavlov, I. V., & Zheltov, I. T. (2005), *Том 2. Otechestvennye bronirovannye mashiny. 1941-1945 gg.* [Volume 2. Domestic Armored Vehicles. 1941-1945.]. M.: Exprint. 344 p.: il. Page 293.
15. *Test Works on Self-Propelled Guns 1943.* TsAMO (Central Archive of the Ministry of Defense of the Russian Federation), F.38, O. 11369, D. 92, Page 7.
16. Ермолов, А. Ю. (2009). Танковая промышленность СССР в годы Великой Отечественной войны. [Tank Industry of the USSR during the Great Patriotic War]. М.: 310 с. Pages 286–287.
17. Мельников, Н. Н. (2014). "Growth in the production of tanks and self-propelled guns in the USSR during the Great Patriotic War." In: *History of Science and Technology in the Modern System of Knowledge: Fourth Annual Conference of the Department of the History of Science and Technology, February 8, 2014.* Yekaterinburg: [Publishing House of UMC UPI]. Page 184.

18. Solyankin, A. G., Pavlov, M. V., Pavlov, I. V., & Zheltov, I. T. (2005), *Tom 2. Otechestvennye bronirovannye mashiny. 1941-1945 gg.* [Volume 2. Domestic Armored Vehicles. 1941-1945.]. M.: Exprint. 344 p.: il. Page 291
19. Ibid. Page 291.
20. October 1, 1945. Production of self-propelled guns categorized by type. TsAMO F.38, O.11369, D.662, Page 02
21. November 22, 1943. Report on inspection of self-propelled artillery regiments of the Western Front, Lieutenant Colonel-Engineer Kostsov. TsAMO F.38, O. 11369, D 273, Page 162.
22. Ibid. Page 168
23. Brief Conclusions Drawn from the Actions of Self-Propelled Artillery Regiments on the Central Front. TsAMO F.062, O. 0000321, D. 0032, Page. 5.
24. Originally, Demyanenko mentions two different classes: истребители танков (tank destroyers) and танки-истребители (tanks-destroyers). However, further in the report, he designates M9, M10, and M10A1 as tank destroyers, implying that he considered them to belong to the same class of self-propelled guns
25. August 14, 1943. Lieutenant Colonel-Engineer Demyanenko. "American Tank Industry," thesis for report. TsAMO F. 38, O. 11369, D. 428, Pages. 53–54.
26. July 15, 1945. Alymov, N. N. Report from the Chief of the Department of Self-Propelled Artillery to the Chief of the Main Armored Directorate of the Red Army on the Prospects of the Development of Self-Propelled Artillery of the Red Army. TsAMO F. 38, O. 11369, D. 698, Pages 36–41.
27. Ibid. Pages 36–41.

Chapter 2

28. Яковлев, Н.Д. Об артиллерии и немного о себе. [Yakovlev, N.D. *On Artillery and a Little About Myself*]. М.: Высшая школа, 1984.
29. Ibid.
30. December 21, 1942. г. Приказ о сформировании управления механической тяги и самоходной артиллерии в составе Главного артиллерийского управления Красной Армии. [Order on the Formation of the Directorate of Mechanical Traction and Self-Propelled Artillery within the Main Artillery Directorate of the Red Army]. RGVA, F. 4, O. 11, D. 73, Pages 377–379.
31. Ibid.
32. April 20, 1945. Report of the UFiU on the Development of Tanks and Armored Vehicles during the Great Patriotic War. TsAMO F. 38, O. 11373, D. 158, Page 36.
33. April 23, 1943. On the Transfer of Self-Propelled Artillery under the Command of the Commander of the Red Army's Armored and Mechanized Forces. RGVA, F. 4, O. 11, D. 75, Pages 588–589. Original Document.
34. Яковлев, Н.Д., Об артиллерии и немного о себе. [Yakovlev, N.D., *On Artillery and a Little About Myself*]. М.: Высшая школа, 1984.
35. Отчеты танковых военных лагерей о проделанной работе за *1941-1945*. [Reports of Tank Military Camps on the Work Done during 1941-1945]. TsAMO F. 38, O. 11373, D. 157, Page 2.
36. Начальствующий состав. (n.d.). Военный энциклопедический словарь [Military Encyclopedical Dictionary]. Retrieved from https://encyclopedia.mil.ru/encyclopedia/dictionary/details.htm?id=6942@morfDictionary
37. Отчеты танковых военных лагерей о проделанной работе за *1941-1945*. [Reports of Tank Military Camps on the Work Done during 1941-1945]. TsAMO F. 38, O. 11373, D. 157, Page 2.
38. Ibid.
39. Ibid. Page 4.
40. Ibid.
41. Ibid.
42. August 16, 1945. Ведомость о количестве сформированных, доукомплектованных и переформированных танковых и механизированных соединений и частей, а также

соединений и частей самоходной артиллерии с августа *1941* по май *1945*. [Statement on the Number of Formed, Replenished, and Reformed Tank and Mechanized Units and Formations, as well as Units and Formations of Self-Propelled Artillery from August 1941 to May 1945]. TsAMO F. 38, O. 11355, D. 981, Page 20.
43. Отчеты танковых военных лагерей о проделанной работе за 1941-1945. [Reports of Tank Military Camps on the Work Done during 1941-1945]. TsAMO F. 38, O. 11373, D. 157, Page 6.
44. July 5. 1945. Отчет штаба бронетанковых и механизированных войск Красной Армии за 1944 год. [Report of the Headquarters of the Red Army's Armored and Mechanized Forces for the year 1944]. TsAMO F. 38, O. 11355, D. 979, Pages 19–20.
45. August 16, 1945. Ведомость о количестве сформированных, доукомплектованных и переформированных танковых и механизированных соединений и частей, а также соединений и частей самоходной артиллерии с августа 1941 по май 1945. [Statement on the Number of Formed, Replenished, and Reformed Tank and Mechanized Units and Formations, as well as Units and Formations of Self-Propelled Artillery from August 1941 to May 1945]. TsAMO F. 38, O. 11355, D. 981, Pages 19–20.
46. Использование самоходной артиллерии в отечественной войне. [The Use of Self-Propelled Artillery in the Great Patriotic War]. TsAMO F. 236, O. 2673, D. 2008, Pages 41–42.
47. Отчеты танковых военных лагерей о проделанной работе за 1941-1945. [Reports of Tank Military Camps on the Work Done during 1941-1945]. TsAMO F. 38, O. 11373, D. 158, Page 42.
48. January 4, 1943, Order #20, "On Strengthening the Firepower of Tank and Mechanized Units and Formations of the Red Army." RGVA, F. 4, O. 11, D. 75, Page 32.
49. April 20, 1945. Report of the UFiU on the Development of Tanks and Mechanized Forces during the Great Patriotic War. TsAMO F.38, O. 11373, D. 158, Pages 34–42
50. January 3, 1944. На основании директивы ГУФ и БП БТ и МВ КА № 1118493 от 30 декабря 1943 г. бригада укомплектовывается. [Based on the Directive of the Main Directorate of Armored Forces and the Directorate of Tank and Motorized Forces of the Red Army No. 1118493 dated December 30, 1943, the brigade is being equipped.] TsAMO, F. 3308, O. 0000001, D. 0018, Page 1.
51. Отчеты танковых военных лагерей о проделанной работе за 1941-1945. [Reports of Tank Military Camps on the Work Done during 1941-1945]. TsAMO F. 38, O. 11373, D. 158, Pages 42–45.
52. Ibid.
53. Ibid. Page 46.
54. Ibid. Page 45.

Chapter 3
54. April 23, 1943. On the Transfer of Self-Propelled Artillery to the Command of the Red Army Armored and Mechanized Forces. RGVA F. 4, O. 11, D. 75, Pages 588–589. Original Document.
55. Отчеты танковых военных лагерей о проделанной работе за 1941-1945. [Reports of Tank Military Camps on the Work Done during 1941-1945]. TsAMO F. 38, O. 11373, D. 157, Page 6.
56. Ibid.
57. Ibid. Page 34.
58. Ibid. Page 4.
59. Ibid. Page 17.
60. Ibid.
61. Ibid.
62. Ibid.
63. Ibid. Page 47.
64. Ibid. Page 18.
65. Ibid.
66. Ibid. Page 4.

67. Ibid. Page 17.
68. Ibid. Page 18.
69. Ibid.
70. Ibid.
71. Ibid. Page 48.
72. Ibid. Page 39.
73. Ibid. Page 157.
74. Ibid. Page 48.
75. July 5. 1945. Отчет штаба бронетанковых и механизированных войск Красной Армии за 1944 год. [Report of the Headquarters of the Red Army's Armored and Mechanized Forces for the year 1944]. TsAMO F. 38, O. 11355, D. 979, Page 32.
76. Ibid. Page 34.
77. April 16, 1946. Report on the work of the Department of Formation and the Directorate of Armored and Mechanized Troops of the Red Army during the period of the Great Patriotic War. TsAMO F. 38, O. 11373, D. 152, Pages 39–40.
78. October 16, 1943. Order No. 0436 of the People's Commissar of Defense of the USSR on the transfer of training tank units for a 6-month training period. TsAMO F2, O. 493437, D.11, p. 625.
79. Отчеты танковых военных лагерей о проделанной работе за 1941-1945. Reports of Tank Military Camps on the Work Done during 1941-1945. TsAMO F. 38, O. 11373, D. 157, Pages 39–40.
80. Ibid. Pages 197–198.
81. Ibid. Pages 217–218.
82. Ibid. Page 37.
83. Ibid. Page 15.
84. Ibid. Pages 146–147.
85. Ibid. Pages 218–219.
86. June 8, 1944. Report on the combat training of units of the 12th Self-Propelled Artillery Brigade. TsAMO, F. 3304, O. 0000001, D. 0045, Page 8.
87. Ibid. Pages 8–9.
88. April 30, 1945. Report on the combat training of the 377th Guards Self-Propelled Artillery Regiment. TsAMO F 4315, O. 0484638, D. 0002, Pages 58–59.
89. November 20, 1944. Fedorenko, Y., & Biryukov, N. On the preparation of armored and mechanized units in 1944. TsAMO F. 38, O. 11339, D 5, Pages 121–123. Copy.
90. Ibid
91. Ibid
92. Ibid
93. November 22, 1944. Fedorenko, Y., & Biryukov, N. Directive of the Commander of the Tank and Mechanized Forces on the Improvement of Training for Tank and Self-Propelled Units. TsAMO. F. 38, O. 11353, D. 199, Pages 217–218. Copy.

Chapter 4

94. *Collection of Combat Documents of the Great Patriotic War.* Issue 15. Moscow: Military Publishing, 1952. Pages 98-102.
95. January 5, 1943. "Temporary Regulations for the Combat Employment of Self-Propelled Artillery." Cited in: *Collection of Combat Documents of the Great Patriotic War.* Issue 15. Moscow: Military Publishing, 1952. Page 98.
96. Ibid. Page 100.
97. Ibid. Page 99.
98. Ibid. Page 98.
99. December 1931. Triandafillov, V. "Deep Tactics." RGVA F.4, O.14, D.760, Page 7.
100. *The Great Soviet Encyclopedia, 3rd Edition.* S.v. "Roving Guns." Retrieved December 7 2023 from https://encyclopedia2.thefreedictionary.com/Roving+Guns

101. January 5, 1943. "Temporary Regulations for the Combat Employment of Self-Propelled Artillery." Cited in: *Collection of Combat Documents of the Great Patriotic War*. Issue 15. Moscow: Military Publishing, 1952. Page 101.
102. Ibid. Page 98.
103. July 25, 1943. Combat Order No. 00863 of the Commander of the Bryansk Front regarding the combat employment of self-propelled artillery. AMO SSSR F. 659, O. 22043c, D. 2, Page 62.
104. January 6, 1944, Combat Order No.03 of the Commander of the Tank and Mechanized Forces of the Belorussian Front "On the combat employment of self-propelled artillery." AMO SSSR F. 426, O. 95691c, D. 1, Pages 6–8.
105. Combat Regulations of the Red Army's Armored and Mechanized Forces. Part 1, Moscow, 1944, Page 4.
106. January 6, 1944, Combat Order No.03 of the Commander of the Tank and Mechanized Forces of the Belorussian Front "On the combat employment of self-propelled artillery." AMO SSSR F. 426, O. 95691c, D. 1, Pages 6–8.
107. September, 1944. Kurgalin A.F, Domnich A.I. "Combat manual on employment and actions for self-propelled artillery," Military Publishing of the People's Commissariat of Defense, Moscow. Page 4.
108. Ibid.
109. December 12, 1944. Memorandum for the Self-Propelled Artillery Crew, Directorate of Tank and Mechanized Forces of the 1st Belorussian Front. Cited in: *Collection of Combat Documents of the Great Patriotic War*. Issue 15. Moscow: Military Publishing, 1952. Page 105.
110. September, 1944. Kurgalin A.F, Domnich A.I. "Combat manual on employment and actions for self-propelled artillery," Military Publishing of the People's Commissariat of Defense, Moscow. Page 5.
111. May 29, 1944. Instructions for the Combat Use of IS-122 Tank Regiments and ISU-152 Self-Propelled Artillery Regiments. Cited in: *Collection of Combat Documents of the Great Patriotic War*. Issue 2. Moscow: Military Publishing, 1947. Page 69.
112. July 25, 1943. Combat Order No. 00863 of the Commander of the Bryansk Front regarding the combat employment of self-propelled artillery. AMO SSSR F. 659, O. 22043c, D. 2, Page 62.
113. January 6, 1944, Combat Order No.03 of the Commander of the Tank and Mechanized Forces of the Belorussian Front "On the combat employment of self-propelled artillery." AMO SSSR F. 426, O. 95691c, D. 1, Pages 6–8.
114. Ibid.
115. Ibid.
116. November 17, 1944. Directive of the Commander of the 1st Baltic Front No. 0130342 of November 17, 1944, on deficiencies in the use of self-propelled artillery operating jointly with infantry units and measures for their elimination. AMO SSSR F.407, O. 57681c, D.4, Pages 217–220.
117. Ibid.
118. January 5, 1943. "Temporary Regulations for the Combat Employment of Self-Propelled Artillery." Cited in: *Collection of Combat Documents of the Great Patriotic War*. Issue 15. Moscow: Military Publishing, 1952. Page 98.
119. January 6, 1944, Combat Order No.03 of the Commander of the Tank and Mechanized Forces of the Belorussian Front "On the combat employment of self-propelled artillery." AMO SSSR F. 426, O. 95691c, D. 1, Pages 6–8.
120. June 12, 1944. Fedorenko, Y., & Biryukov, N. Directive of the Commander of the Tank and Mechanized Forces on the Combat Employment of Heavy Tank and Self-Propelled Artillery Regiments. TsAMO. F. 38, O. 11353, D. 199, Pages 150–153. Copy.
121. January 5, 1943. "Temporary Regulations for the Combat Employment of Self-Propelled Artillery." Cited in: *Collection of Combat Documents of the Great Patriotic War*. Issue 15. Moscow: Military Publishing, 1952. Page 105.
122. November 22, 1943. Report on inspection of self-propelled artillery regiments of the Western Front, Lieutenant Colonel-Engineer Kostsov. TsAMO F.38, O. 11369, D 273, Pages 161–162.

123. Trials and Employment of Self-Propelled Guns in 1944. TsAMO F.38, O. 11369, D. 273, Page 54.
124. The Use of Self-Propelled Artillery in the Great Patriotic War. TsAMO F. 236, O. 2673, D. 2008, *Page 41*
125. June 16, 1945. Report on the Combat Use of Tank Armament in the Great Patriotic War 1941-1945. 367th Guards Self-Propelled Artillery Regiment. TsAMO F.3421, Op. 0000001, D. 0038, Pages 17–19.
126. June 27, 1945. "On the Employment of Tank (Self-Propelled) Weapons in the Great Patriotic War." TsAMO F.323, O. 4756, D. 169, Page 178.
127. Ibid.
128. January 6, 1944, Combat Order No.03 of the Commander of the Tank and Mechanized Forces of the Belorussian Front "On the combat employment of self-propelled artillery." AMO SSSR F. 426, O. 95691c, D. 1, Pages 6–8.
129. November 22, 1943. Report on inspection of self-propelled artillery regiments of the Western Front, Lieutenant Colonel-Engineer Kostsov. TsAMO F.38, O. 11369, D 273, Pages 161–172.
130. November 17, 1944. Directive of the Commander of the 1st Baltic Front No. 0130342 of November 17, 1944, on deficiencies in the use of self-propelled artillery operating jointly with infantry units and measures for their elimination. AMO SSSR F.407, O. 57681c, D.4, Pages 217–220.
131. January 5, 1943. "Temporary Regulations for the Combat Employment of Self-Propelled Artillery." Cited in: *Collection of Combat Documents of the Great Patriotic War*. Issue 15. Moscow: Military Publishing, 1952. Page 104.
132. January 6, 1944, Combat Order No.03 of the Commander of the Tank and Mechanized Forces of the Belorussian Front "On the combat employment of self-propelled artillery." AMO SSSR F. 426, O. 95691c, D. 1, Pages 6–8.
133. November 17, 1944. Directive of the Commander of the 1st Baltic Front No. 0130342 of November 17, 1944, on deficiencies in the use of self-propelled artillery operating jointly with infantry units and measures for their elimination. AMO SSSR F.407, O. 57681c, D.4, Pages 217–220.
134. "Guidelines for the Combat Use of Self-Propelled Artillery Regiments (SU-76) and their Tactical Cooperation with Infantry. 1945." Cited in: *Collection of Combat Documents of the Great Patriotic War*. Issue 2. Moscow: Military Publishing, 1947. Pages 76–78.
135. June 27, 1945. "On the Employment of Tank (Self-Propelled) Weapons in the Great Patriotic War." TsAMO F.323, O. 4756, D169, Page 177.
136. June 12, 1944. Fedorenko, Y., & Biryukov, N. Directive of the Commander of the Tank and Mechanized Forces on the Combat Employment of Heavy Tank and Self-Propelled Artillery Regiments. TsAMO. F. 38, O. 11353, D. 199, Pages 150–153. Copy.
137. Ibid.
138. May 29, 1944. Novikov, Petrov. "Guidelines for the Combat Use of IS-122 Tank Regiments and ISU-152 Self-Propelled Artillery Regiments." Page 71. Cited in: *Collection of Combat Documents of the Great Patriotic War*. Issue 2. Moscow: Military Publishing, 1947. Pages 67–71.
139. June 12, 1944. Fedorenko, Y., & Biryukov, N. Directive of the Commander of the Tank and Mechanized Forces on the Combat Employment of Heavy Tank and Self-Propelled Artillery Regiments. TsAMO. F. 38, O. 11353, D. 199, Pages 150–153. Copy.
140. May 29, 1944. Novikov, Petrov. "Guidelines for the Combat Use of IS-122 Tank Regiments and ISU-152 Self-Propelled Artillery Regiments." Page 67–69. Cited in: *Collection of Combat Documents of the Great Patriotic War*. Issue 2. Moscow: Military Publishing, 1947. Pages 67–71.
141. June 12, 1944. Fedorenko, Y., & Biryukov, N. Directive of the Commander of the Tank and Mechanized Forces on the Combat Employment of Heavy Tank and Self-Propelled Artillery Regiments. TsAMO. F. 38, O. 11353, D. 199, Pages 150–153. Copy.
142. May 29, 1944. Novikov, Petrov. "Guidelines for the Combat Use of IS-122 Tank Regiments and ISU-152 Self-Propelled Artillery Regiments." Page 67–69. Cited in: *Collection of Combat Documents of the Great Patriotic War*. Issue 2. Moscow: Military Publishing, 1947. Pages 67–71.

143. June 12, 1944. Fedorenko, Y., & Biryukov, N. Directive of the Commander of the Tank and Mechanized Forces on the Combat Employment of Heavy Tank and Self-Propelled Artillery Regiments. TsAMO. F. 38, O. 11353, D. 199, Pages 150–153. Copy.
144. May 29, 1944. Novikov, Petrov. "Guidelines for the Combat Use of IS-122 Tank Regiments and ISU-152 Self-Propelled Artillery Regiments." Page 71. Cited in: *Collection of Combat Documents of the Great Patriotic War*. Issue 2. Moscow: Military Publishing, 1947. Pages 67–71.
145. Ibid.
146. Zaloga, S. J. (n.d.). *IS-2 Heavy Tank 1944–73*. Osprey Publishing. Page 11.
147. June 20, 1944. "Employment of Heavy "IS" Tank Regiments in the 11th Guards Tank Corps." TsAMO F. 236, O. 2673, D. 299, Pages 115–118.
148. January 6, 1944, Combat Order No.03 of the Commander of the Tank and Mechanized Forces of the Belorussian Front "On the combat employment of self-propelled artillery." AMO SSSR F. 426, O. 95691c, D. 1, Pages 6–8.
149. September, 1944. Kurgalin A.F, Domnich A.I. "Combat manual on employment and actions for self-propelled artillery," Military Publishing of the People's Commissariat of Defense, Moscow. Pages 68–69.
150. Ibid.
151. April 25, 1945. 'Self-propelled Artillery in Urban Combat." TsAMO F.4334, O. 0184441c, D. 0002, Pages 164–165.
152. Ibid.
153. Ibid.
154. June 16, 1945. Report on the Combat Use of Tank Armament in the Great Patriotic War 1941-1945. 367th Guards *Self-Propelled Artillery* Regiment. TsAMO F.3421, Op. 0000001, D. 0038, Pages 17–19.
155. Employment of Self-Propelled Artillery in the Great Patriotic War. TsAMO, F: 236, O. 2673, D. 2008, Page 311.

Chapter 5

156. August 20, 1943. Brief Conclusions Drawn from the Actions of Self-Propelled Artillery Regiments on the Central Front. TsAMO F.062, O. 0000321, D. 0032, Pages 1–5.
157. January 5, 1943."Temporary Regulations for the Combat Employment of Self-Propelled Artillery." Cited in: *Collection of Combat Documents of the Great Patriotic War*. Issue 15. Moscow: Military Publishing, 1952. Page 98.
158. July 25, 1943. Combat Order No. 00863 of the Commander of the Bryansk Front regarding the combat employment of self-propelled artillery. AMO SSSR F. 659, O. 22043c, D. 2, Page 62.
159. August 20, 1943. Brief Conclusions Drawn from the Actions of Self-Propelled Artillery Regiments on the Central Front. TsAMO F.062, O. 0000321, D. 0032, Pages 1–5.
160. January 6, 1944, Combat Order No.03 of the Commander of the Tank and Mechanized Forces of the Belorussian Front "On the combat employment of self-propelled artillery." AMO SSSR F. 426, O. 95691c, D. 1, Pages 6–8.
161. *November 22, 1943. Report on inspection of self-propelled artillery regiments of the Western Front, Lieutenant Colonel-Engineer Kostsov*. TsAMO F.38, O. 11369, D 273, Pages 161–162.
162. Ibid. Page 166.
163. January 6, 1944, Combat Order No.03 of the Commander of the Tank and Mechanized Forces of the Belorussian Front "On the combat employment of self-propelled artillery." AMO SSSR F. 426, O. 95691c, D. 1, Pages 6–8.
164. November 17, 1944. Directive of the Commander of the 1st Baltic Front No. 0130342 of November 17, 1944, on deficiencies in the use of self-propelled artillery operating jointly with infantry units and measures for their elimination. AMO SSSR F.407, O. 57681c, D.4, Pages 217–220.
165. March 3, 1945. "Deficiencies in the Combat Use of Towed and Self-Propelled Artillery in the Operations Conducted by the 5th Guards Tank Army from January 17 to February 28, 1945."

Cited in: *Collection of Combat Documents of the Great Patriotic War*. Issue 15. Moscow: Military Publishing, 1952. Pages 104–105.
166. Ibid.
167. Ibid. Page 104.
168. December 17, 1945. "Self-Propelled Artillery of the Tank Corps: Role and Tasks, Place in Combat Formation, and Reinforcement Standards for Tank Units." TsAMO F. 3421, O. 0000001, D. 0038, Pages. 128-129 with rev.
169. December 15, 1945. Theses of the report on the combat experience of artillery support during the actions of a tank corps in the operational depth, presented at the army conference of the 5th Guards Tank Army, which took place from December 13 to 15, 1945, for the purpose of studying, summarizing, and utilizing the experience of the Great Patriotic War. TsAMO F. 332, O. 4948, D: 441, Page 330.
170. December 17, 1945. "Self-Propelled Artillery of the Tank Corps: Role and Tasks, Place in Combat Formation, and Reinforcement Standards for Tank Units." TsAMO F. 3421, O. 0000001, D. 0038, Pages. 128–129 with rev.
171. *November 22, 1943. Report on inspection of self-propelled artillery regiments of the Western Front, Lieutenant Colonel-Engineer Kostsov.* TsAMO F.38, O. 11369, D 273, Pages 161, 167.
172. October 31, 1943. "Report on the Combat Actions of Self-Propelled Artillery Regiments of the 1st Ukrainian Front for October 1943." TsAMO F. 236, O. 2673, D. 130, Pages 148, 152–153.
173. September, 1944. Kurgalin A.F, Domnich A.I. "Combat manual on employment and actions for self-propelled artillery," Military Publishing of the People's Commissariat of Defense, Moscow. Page 49.
174. August 14, 1945. "A Report Summarizing the Combat Experience of the Great Patriotic War in the Use of Self-Propelled Artillery and the Support of Tank and Mechanized Corps Breakthrough." TsAMO F. 332, O. 4948, D. 440, Page 19.
175. September 9, 1945. "1051st Self-Propelled Artillery Regiment's War Diary." TsAMO F. 4390, O. 0342384c, D. 0001, Page 7.
176. November 17, 1944. Directive of the Commander of the 1st Baltic Front No. 0130342 of November 17, 1944, on deficiencies in the use of self-propelled artillery operating jointly with infantry units and measures for their elimination. AMO SSSR F.407, O. 57681c, D.4, Pages 217–220.
177. May 9, 1945. "War Diary of the 29th Tank Corps." TsAMO F. 3420, O. 1, D. 25, Pages 47–49.
178. Ibid.
179. August 14, 1945. "A Report Summarizing the Combat Experience of the Great Patriotic War in the Use of Self-Propelled Artillery and the Support of Tank and Mechanized Corps Breakthrough." TsAMO F. 332, O. 4948, D. 440, Pages 19–22.
180. May 9, 1945. "War Diary of the 29th Tank Corps." TsAMO F. 3420, O. 1, D. 25, Pages 59–64.
181. February 10, 1945. Report on the Combat Actions of the 31st Tank Brigade in Northern Poland and Eastern Prussia. TsAMO 3114, O. 0000001, D. 0014, Page 12.
182. Ibid. Pages 2–3.
183. August 14, 1945. "A Report Summarizing the Combat Experience of the Great Patriotic War in the Use of Self-Propelled Artillery and the Support of Tank and Mechanized Corps Breakthrough." TsAMO F. 332, O. 4948, D. 440, Pages 21–22.
184. Tarasov, A. (2021), *Red Army Auxiliary Armoured Vehicles 1930-1945*. Pen & Sword. Pages 14–15.
185. June 5, 1929. Triandafillov V.K "On the system of tank-tractor-auto-armoured weapons of RKKA." RGVA, F. 4, O. 2, D. 504, Pages 5–18 s ob.
186. February 10, 1945. Report on the Combat Actions of the 31st Tank Brigade in Northern Poland and Eastern Prussia. TsAMO 3114, O. 0000001, D. 0014, Page 12.
187. Ibid.
188. August 14, 1945. "A Report Summarizing the Combat Experience of the Great Patriotic War in the Use of Self-Propelled Artillery and the Support of Tank and Mechanized Corps Breakthrough." TsAMO F. 332, O. 4948, D. 440, Page 21.

189. Ibid. Page 19.
190. Ibid. Pages 20–21.
191. Ibid. Page 22.
192. May 9, 1945. "War Diary of the 29th Tank Corps." TsAMO F. 3420, O. 1, D. 25, Pages 47–49, 80–81.
193. Ibid. Pages 47–49, 57–59, 75–76, 77–78, 80–81.
194. August 15, 1945. "Table of Equipment and Losses of Tank and Mechanized Units of the 2nd Belorussian Front in the East Prussian Operation." TsAMO F.46, Op.2394, D 1547, pp. 181–182.
195. Grau, L. W., & Bartles, C. K. (2018, May). "The Russian Reconnaissance Fire Complex Comes of Age." Changing Character of War Centre, Pembroke College, University of Oxford, with Axel and Margaret Axson Johnson Foundation. Page 1.

Chapter 6

196. Solyankin, A. G., Pavlov, M. V., Pavlov, I. V., & Zheltov, I. T. (2005), *Tom 2. Otechestvennye bronirovannye mashiny. 1941-1945 gg.* [Volume 2. Domestic Armored Vehicles. 1941-1945.]. M.: Exprint. 344 p.: il. Page 361.
197. Hunnicutt, R. P. (n.d.), *Half-Track: A History of American Semi-Tracked Vehicles*. Echo Point Books & Media. Page 109.
198. June 4, 1945. "Report on the Receipt of Imported Tanks." TsAMO F. 38, O. 11355, D. 3016, Page 118.
199. April 12, 1944. "Report on the Testing of the American T-48 Self-Propelled Gun." TsAMO F. 38, O. 11369, D. 363, Page 36.
200. Ibid.
201. May 9, 1945. "On the Combat Use of the Self-Propelled Brigade SU-57-I /From the Experience of Combat Actions of the 16th Guards Self-Propelled Artillery Brigade/." TsAMO F. 3306, O. 0000001, D. 0029, Page 83.
202. July 5, 1945. Отчет штаба бронетанковых и механизированных войск Красной Армии за 1944 год. [Report of the Headquarters of the Red Army's Armored and Mechanized Forces for the year 1944]. TsAMO F. 38, O. 11355, D. 979, Pages 43, 44-2.
203. November 17, 1943. "Report on the Testing of the American Self-Propelled Gun M-10." TsAMO F. 38, O. 11369, D. 51, Page 37.
204. Ibid.
205. May 23, 1944. "Report on the Testing of the American Self-Propelled Gun T-70." TsAMO F. 38, O. 11355, D. 2387, Pages 2–6.
206. October 23, 1944. "Report on the Testing of the American Self-Propelled Gun M15-A1." TsAMO F.38, O.11369, D.362, Page 1.
207. Bystrova, I. V. (2019), Ленд-лиз для СССР: экономика, техника, люди (1941–1945). Page 237.
208. Ibid.
209. June 4, 1945. "Report on the Receipt of Imported Tanks." TsAMO F. 38, O. 11355, D. 3016, Pages 118, 185.
210. August 4, 1944. "Report on the Combat Actions of the 1123rd Light Self-Propelled Artillery Regiment from June 25, 1944, to July 30, 1944." TsAMO F. 4412, O. 0088450c, D. 0002, Page 7.
211. Ibid.
212. April 20, 1945. Report of the UFiU on the Development of Tanks and Armored Vehicles during the Great Patriotic War. TsAMO F. 38, O. 11373, D. 158, Page 42.
213. February 28, 1945. "Material for Analysis of Combat Actions Conducted by the 16th Guards Self-Propelled Artillery Brigade in January-February 1945." TsAMO F. 3306, Op: 0000001, D: 0029, Page 52.
214. May 9, 1945. "On the Combat Use of the Self-Propelled Brigade SU-57-I /From the Experience of Combat Actions of the 16th Guards Self-Propelled Artillery Brigade/." TsAMO F. 3306, O. 0000001, D. 0029, Page 83, 87.

215. February 28, 1945. "Material for Analysis of Combat Actions Conducted by the 16th Guards Self-Propelled Artillery Brigade in January-February 1945." TsAMO F. 3306, Op: 0000001, D: 0029, Page 52.
216. Bystrova, I. V. (2019), Ленд-лиз для СССР: экономика, техника, люди (1941–1945). Page 237.
217. June 27, 1945. "On the Employment of Tank (Self-Propelled) Weapons in the Great Patriotic War." TsAMO F.323, O. 4756, D. 169, Page 237.
218. Ibid.
219. July 16, 1945. "Materials on Generalizing the Experience of the Great Patriotic War in the Use of Armored Cars and Armored Personnel Carriers." TsAMO F. 3138, O. 0000001, D. 0009, Page 86.
220. Ibid.
221. July 16, 1945. "Responses to Questions on Armored Cars and Armored Personnel Carriers." TsAMO F. 323, O. 4756, D. 169, Page 251.
222. July 13, 1945. "Generalized Experience of the Use of Armored Personnel Carriers and Armored Vehicles by Units of the 48th Army, as well as the Employment of Enemy Armored Personnel Carriers and Armored Vehicles on the Front of the Army in the Great Patriotic War." TsAMO F. 446, O. 9657, D. 490, Page 135.
223. Bystrova, I. V. (2019). Ленд-лиз для СССР: экономика, техника, люди (1941–1945). Page 237.
224. Ibid.
225. Ibid. Pages 431–432.
226. June 4, 1945. "Report on the Receipt of Imported Tanks." TsAMO F. 38, O. 11355, D. 3016, Pages 118, 185.
227. October 4, 1944. "Report on the trials of the American anti-aircraft self-propelled gun M-17." TsAMO F 38, O. 11369, D. 341, Page 1.

Conclusion

228. Solyankin, A. G., Pavlov, M. V., Pavlov, I. V., & Zheltov, I. T. (2005), *Tom 2. Otechestvennye bronirovannye mashiny. 1941-1945 gg.* [Volume 2. Domestic Armored Vehicles. 1941-1945.]. M.: Exprint. 344 p.: il. Pages 10-12.
229. Melnikov, N. N. "Growth of Tank and Self-Propelled Artillery Production in the USSR During the Great Patriotic War." Page 184.
230. Ermolov, A. (2009). "Tank Industry of the USSR during the Great Patriotic War." Moscow. The table was compiled according to: RGAE. F. 8756. O. 4. D. 728. Pages 158–164.
231. Ibid. Page 87.
232. July 13, 1945. "Report of the UFiU of Tanks and Mechanized Forces for the Period of the Great Patriotic War." TsAMO F. 38, O. 11373, D. 158, Pages 53–54.
233. Ibid. Page 54.
234. Ibid.
235. Department of the Army. (2012, May 16). ADRP 3-0: Unified Land Operations. Pages 1-14.
236. February 10, 1945. Report on the Combat Actions of the 31st Tank Brigade in Northern Poland and Eastern Prussia. TsAMO 3114, O. 0000001, D. 0014, Pages 21–22.
237. November 22, 1943. Report on inspection of self-propelled artillery regiments of the Western Front, Lieutenant Colonel-Engineer Kostsov. TsAMO F.38, O. 11369, D 273, Page 163.
238. August 14, 1945. "A Report Summarizing the Combat Experience of the Great Patriotic War in the Use of Self-Propelled Artillery and the Support of Tank and Mechanized Corps Breakthrough." TsAMO F. 332, O. 4948, D. 440, Pages 19–22.
239. V.I. Feskov, V.I. Golikov, K.A. Kalashnikov, S.A. Slugin (2013), *The Armed Forces of the USSR after the Second World War: From the Red Army to the Soviet. Part 1. Ground Forces*. Tomsk, Izdatelstro NTL. Page 273.

Dear Reader,

We hope you have enjoyed this book, but why not share your views on social media? You can also follow our pages to see more about our other products: facebook.com/penandswordbooks or follow us on X @penswordbooks

You can also view our products at www.pen-and-sword.co.uk (UK and ROW) or www.penandswordbooks.com (North America).

To keep up to date with our latest releases and online catalogues, please sign up to our newsletter at: www.pen-and-sword.co.uk/newsletter

If you would like a printed catalogue with our latest books, then please email: enquiries@pen-and-sword.co.uk or telephone: 01226 734555 (UK and ROW) or email: uspen-and-sword@casematepublishers.com or telephone: (610) 853-9131 (North America).

We respect your privacy and we will only use personal information to send you information about our products.

Thank you!